SpringerBriefs in Philosophy

SpringerBriefs present concise summaries of cutting-edge research and practical applications across a wide spectrum of fields. Featuring compact volumes of 50 to 125 pages, the series covers a range of content from professional to academic. Typical topics might include:

- A timely report of state-of-the art analytical techniques
- A bridge between new research results, as published in journal articles, and a contextual literature review
- A snapshot of a hot or emerging topic
- An in-depth case study or clinical example
- A presentation of core concepts that students must understand in order to make independent contributions

SpringerBriefs in Philosophy cover a broad range of philosophical fields including: Philosophy of Science, Logic, Non-Western Thinking and Western Philosophy. We also consider biographies, full or partial, of key thinkers and pioneers.

SpringerBriefs are characterized by fast, global electronic dissemination, standard publishing contracts, standardized manuscript preparation and formatting guidelines, and expedited production schedules. Both solicited and unsolicited manuscripts are considered for publication in the SpringerBriefs in Philosophy series. Potential authors are warmly invited to complete and submit the Briefs Author Proposal form. All projects will be submitted to editorial review by external advisors.

SpringerBriefs are characterized by expedited production schedules with the aim for publication 8 to 12 weeks after acceptance and fast, global electronic dissemination through our online platform SpringerLink. The standard concise author contracts guarantee that

- an individual ISBN is assigned to each manuscript
- each manuscript is copyrighted in the name of the author
- the author retains the right to post the pre-publication version on his/her website or that of his/her institution.

More information about this series at https://link.springer.com/bookseries/10082

Christophe Geudens • Lorenz Demey

The Modal Logic of John Fabri of Valenciennes (c. 1500)

A Study in Token-Based Semantics

Springer

Christophe Geudens
FWO/Research Foundation Flanders
KU Leuven – Institute of Philosophy
Leuven, Belgium

Lorenz Demey
KU Leuven – Institute of Philosophy
Leuven, Belgium

Fonds Wetenschappelijk Onderzoek
Research funded by FWO – Research Foundation Flanders (grants 1160017N and 12U1521N).

ISSN 2211-4548 ISSN 2211-4556 (electronic)
SpringerBriefs in Philosophy
ISBN 978-3-030-98801-2 ISBN 978-3-030-98802-9 (eBook)
https://doi.org/10.1007/978-3-030-98802-9

This Springer imprint is published by the registered company Springer Nature Switzerland AG
The registered company address is: Gewerbestrasse 11, 6330 Cham, Switzerland

Contents

Chapter 1
Introduction

Abstract This book is a study of approaches to modal logic at the eve of the Northern Renaissance. At this point in time, the logical discourse in Northern Europe was organized around the so-called *Wegestreit*, the dichotomy between *via antiqua* (traditionalism) and *via moderna* (modernism). Past research on *Wegestreit* logic has mainly focused on issues related to categorical logic and ontology. If and how the discourse on modal logic was shaped by the traditionalism-modernism split is still an open question, and this book constitutes the first attempt at an answer. It contains a case study of the modal logic of the lesser-known logician John Fabri of Valenciennes, who was active at the turn of the sixteenth century in Louvain, at that moment one of the main intellectual centres in Northern Europe. This first chapter introduces Fabri and sketches the intellectual setting in which he worked.

Keywords *Wegestreit* · Northern Renaissance · *Via antiqua* · *Via moderna* · John Fabri of Valenciennes · University of Louvain

There are two periods in the history of medieval Western philosophy where the concepts of school and school membership were particularly important. First, there is the twelfth century, the period of the Paris debates between the *nominales*, or Peter Abelard (1079–1142) *cum suis*, and the *reales*, a term that grouped the followers of, among others, Adam of Balsham (†1157/1169), Alberic of Paris (fl. c. 1130–1140) and Robert of Melun (c. 1100–1167). Second, there are the fifteenth and early sixteenth centuries, which, at least in Northern Europe, were characterized by the so-called *Wegestreit*, or the quarrels between traditionalists (*antiqui*, representatives of the *via antiqua*) and modernists (*moderni*, representatives of the *via moderna*).

The twelfth-century Paris debates and the *Wegestreit* are similar in more than one respect – one of the main issues at stake in both cases was the status of universals, for instance –, though evidently they were as different as they were alike. One of the respects in which they differed concerns a simple sociological fact. Unlike the twelfth-century debates, the *Wegestreit* was fought out against the backdrop of a philosophical landscape that was dominated by universities. It is widely recognized in the scholarship that one of the most important characteristics of the *Wegestreit* was precisely its embedding in the university context. In many surveys of late-medieval

C. Geudens, L. Demey, *The Modal Logic of John Fabri of Valenciennes (c. 1500)*,
SpringerBriefs in Philosophy, https://doi.org/10.1007/978-3-030-98802-9_1

philosophy, one will read that the University of Vienna was strongly modernist, for instance, while the philosophical climate in Paris was eclectic. One will also read that the University of Cologne was one of the few institutions where the *via antiqua* dominated, and chances are that in the same breath one will find a reference to Louvain. Indeed, ever since the late nineteenth century it has been the received view in the scholarship that the University of Louvain, which was founded in 1425, was a bulwark of traditionalism during the *Wegestreit*. Call this the Traditionalist Thesis (**TT** for short).

Traditionalist Thesis (**TT**): The discourse on logic in Louvain during the *Wegestreit* was strongly traditionalist.[1]

Most studies argue for **TT** based on circumstantial evidence, such as testimonies and administrative sources. For instance, reference is often made to the passage from the *Letter to Martin Van Dorp* of Thomas More (1478–1535) where More incidentally remarks that the logicians from Louvain and Paris "are so thoroughly at odds with each other that they do not even agree on a name, since the former want to be called *reales*, and the latter *nominales*".[2] Further evidence is usually found in the statutes of the Arts faculty, which contain several prohibitions to teach modernist doctrines. An amendment to the first edition of the Arts statutes, dated 1427, specifies that anyone who wished to be appointed head (*regens*) of an Arts college should swear an oath that he would "never teach" the philosophy of John Buridan (c. 1300-c. 1360), Marsilius of Inghen (c. 1340–1396) or William of Ockham (c. 1287–1347), who were the main authorities in the *via moderna*; or, by extension, the philosophy of any of "their followers" (*se nunquam doctrinare Buridanum, Marsilium, Ockam, aut eorum sequaces*).[3] The 1447 version of the Arts statutes contains a ban on Ockhamism as well as Wycliffite thought, which had been condemned at the Council of Constance in 1415, and it also specifies that Aristotle should be interpreted according to the views of Albert the Great (c. 1200–1280), Thomas Aquinas (1225–1274), Giles of Rome (c. 1240–1316), or Averroes (1126–1198), who were all authorities in the *via antiqua*.[4] The next known revision of the Arts statutes, dated 1512, extends this list of *authores probati* to include Henry of Ghent (c. 1217–1293) and Duns Scotus (c. 1265–1308), and, more generally, "all other traditionalist *doctores* of good reputation" (*omnes alii doctores antiqui et famati*).[5] Scholars often also turn to the *Acta* of the Arts faculty, containing the reports of the sessions of the Arts council. Two clusters of reports, one from 1486 and the other from 1497,

---

[1]For some recent formulations of **TT**, see Biard (2010: 673); Hoenen (2003: 21–22). Further references are found in Geudens (2020: 5 [n. 22]). For general surveys of late-medieval philosophy and logic, with attention to the institutional embedding, see Ashworth (2016); Kałuża (1998); Marenbon (2012); Swieżawski and Prokopowicz (1990). Specifically on the universities of Vienna, Paris and Cologne, see resp. Ritter (1921–1927); Kałuża (1988); Meuthen (1988).

[2]Morus, *Epistola ad Dorpium* (Kinney 1963: 24 [ll. 1–2]). Translation by Kinney (1963: 25).

[3]See Reusens (1867: 155 [n. 1]) for the Latin text.

[4]See Baudry (1950: 68) for the Latin text. An English translation is found in Guerlac (1989: 34).

[5]See Ehrle (1925: 292 [n. 1]) for the Latin text.

show that in these years a controversy had arisen about groups of lecturers who had allegedly explained Aristotle in modernist fashion. Both groups were punished with a temporary suspension due to their infringements of the faculty statutes.[6]

Several studies have shown that the statutory provision to teach traditionalist logic did not remain a dead letter, and they provided direct evidence, drawn from the extant Louvain commentaries on Aristotle's *Organon*, in support of **TT**. For instance, Papy has pointed out that the so-called *Commentarii Lovanienses* (1535), a commentary on the *Organon* and Porphyry's *Isagoge* that was issued by Faculty of Arts for use in the logic classes at the different Arts colleges and also the oldest such commentary of Louvain origin to have appeared in print, usually follows either Aquinas or Albert.[7] More recently, Bartocci and Masolini have argued that the same goes for the commentaries of Peter de Rivo (c. 1420–1499), who was one of the protagonists in the quarrel over future contingents that took place in Louvain during the 1460s and 1470s.[8]

By contrast, other studies into the preserved Louvain sources on logic have reached results that are not easily reconcilable with **TT**. Verbeke has pointed out that Martin Van Dorp (1485–1525), a lecturer at the Arts college called "the Lily" (*Lilium*) whom historians of logic mainly remember for his involvement in the publication of the 1515 *editio princeps* of the *De inventione dialectica* of Rudolph Agricola (1444–1485), could quote approvingly from both Ockham and Buridan in his *Introductio facilis ad Aristotelis libros logice intelligendos utilissima* (1512).[9] Furthermore, Bartocci has shown that some fifteenth-century Louvain commentaries on Aristotle's *Topica* contain traces of typically modernist theories.[10] The same point is more substantially argued by Geudens, who has shown that the influence of modernism on Louvain theories of topical logic (or dialectic) is clearly discernible throughout the latter half of the fifteenth century, and well into the 1530s.[11] It is still unclear at this point just how large and far-reaching the impact of modernism on the Louvain logical discourse was. But the very fact that modernist theories did find inroads into Louvain, as the studies just mentioned demonstrate, is significant for our appraisal of the position of Louvain in the *Wegestreit*, as it suggests that **TT** is too simplistic.

This book provides further evidence that **TT** is in need of revision. It does not focus on Louvain theories of topical logic, which have been the main concern of the scholarship to date, but rather on theories of modal logic. The book proposes an in-depth analysis and contextualization of the modal logic of John Fabri of Valenciennes. Fabri was a lecturer at the College of the Lily from 1493 until around 1505,

---

[6] Leuven, Rijksarchief, Fonds Oude Universiteit Leuven, MS 712, ff. 40^v, 43^v, 48^r, 51^r, 145^v.

[7] Papy (1999: 169–170). See also Madeira (2006).

[8] See Bartocci and Masolini (2014). On the Louvain Quarrel, see esp. Baudry (1950); Schabel (1995, 1996).

[9] Verbeke (2013: 237).

[10] Bartocci (2016: 862–863; 2017: 225–232).

[11] See Geudens (2020).

Table 1.1 The contents of **S**

Work commented on	Paraphrases	Questions
Porphyry, *Isagoge*	ff. 15^{r}-22^{v}	ff. 101^{r}-136^{v}
Aristotle, *Categoriae*	ff. 23^{r}-35^{v}	ff. 137^{r}-190^{v}
Aristotle, *De interpretatione* I-II	ff. 36^{r}-47^{v}	ff. 191^{r}-223^{r}
Aristotle, *Analytica priora* I-II	ff. 48^{r}-73^{v}	ff. 224^{r}-257^{v}
Aristotle, *Analytica posteriora* I-II	ff. 74^{r}-99^{v}	ff. 258^{r}-271^{v}
Aristotle, *Topica* I-II	ff. 282^{r}-301^{v}	ff. 328^{r}-356^{r}
Auctoritates Aristotelis on *Topica* III-VIII	[ff. 302^{r}-304^{v}]	
Aristotle, *Sophistici elenchi* I-II	ff. 305^{r}-322^{r}	ff. 357^{r}-371^{v}

and served in several functions in the administration of the Louvain Arts Faculty around the turn of the sixteenth century. His most famous student is presumably Martin Van Dorp, already mentioned.[12] As with many of his contemporaries, Fabri's only textual output is a collection of commentaries on parts of the *corpus Aristotelicum*, a result of his teaching activities in the Lily. These commentaries are known to survive in one single manuscript, which is currently preserved in Saint-Omer (Bibliothèque d'agglomération, MS 609; henceforth **S**).[13] **S** stems from the very first years of the sixteenth century. It is a convolute manuscript, consisting of close textual paraphrases – usually called '*continuationes*' or '*expositiones*' in other Louvain sources from the period – of the *Isagoge* and large parts of the *Organon*, as well as sets of questions (*quaestiones*) on these works. The contents of **S** are displayed in Table 1.1.[14] All texts in **S** are written down by a certain Allardus Tassart, who can be identified as the Louvain student 'Dominus Alardus de Sancto Odomaro Tassart' who enrolled in the Lily in October 1501.[15] **S** itself is the written record of a logic course that was taught in that college, which implies that all commentaries in **S** must have been written down in a short period of time, and probably during the course of one academic year. Since Allardus completed the questions on the *Isagoge* in 1502, all other commentaries in **S** must have been completed in or around the same year.

Fabri's authorship of part of the texts in **S** is certain. Allardus ascribes the questions on the *Isagoge* and *An. post.* I-II, as well as the paraphrase of *An. pr.* I-II, to a certain Johannes de Valenchenis. Geudens has argued elsewhere that there is little doubt this Johannes is to be identified with John Fabri of Valenciennes.[16] Moreover, we know that in the Louvain Arts colleges during the later fifteenth and early sixteenth centuries, the lectures on the *Organon* were usually the responsibility

[12] For biographical data on Fabri, see Geudens (2018: 82–92; 2020: 314–320), on which the below paragraphs on **S** and Fabri's authorship of the commentaries in **S** are based.

[13] The manuscript is freely accessible at https://bibliotheque-numerique.bibliotheque-agglo-stomer.fr/notices/item/1805-recueil-de-commentaires-sur-aristote-et-porphyre

[14] An edition of **S**'s paraphrase of and questions on *Top.* I is found in Geudens (2018).

[15] Schillings (1962: 212 [nr. 172]).

[16] See the references in fn. 12.

of not one but two lecturers: one read the entire *logica vetus* and part of the *logica nova* (usually the *Analytica priora* or *posteriora*), and the other read the remainder of the *logica nova*. In the sources, these lecturers are sometimes called resp. *magistri anteprandiales* and *postprandiales*, after the time – before or after lunch (*prandium*) – at which their lectures took place. That Fabri authored the questions on the *Isagoge* alongside part of the commentaries on both *Analytica*'s suggests he was a *magister anteprandialis*, endowed with the task of reading the *logica vetus* and the first books of the *logica nova*. Thus the paraphrases of the *Isagoge* and *An. post* I-II, the questions on *An. pr.* I-II, and the paraphrases of and questions on the *Categoriae* and *De interpretatione* are almost certainly also of his hand. Who authored the other commentaries in **S**, on the *Topica* and *Sophistici elenchi*, is unknown. Furthermore, Fabri started lecturing in the Lily immediately after he graduated from this college in late 1493. Thus, bearing in mind that **S** stems from c. 1502, Fabri's commentaries on the *logica vetus* and both *Analytica*'s should be dated somewhere between 1493 and 1502.

Aristotle set out his modal logic at *De int.* 12-13 and *An. pr.* I.3 and I.8-22, and the questions in **S** that relate to these chapters allow for a detailed reconstruction of Fabri's own modal logic. The main argument of this book is that, contrary to what we expect based on **TT**, Fabri's modal logic squarely belongs to the same tradition as the modal logics that were developed by adherents of the *via moderna* in Paris, and represents the approach to modal logic that came to maturation in the circle of John Buridan in the fourteenth century. Thus the book casts further doubt on the truth of **TT**, and, importantly, it shows that the influence of modernism on Louvain logicians goes well beyond the case of topical logic, which, as indicated, has enjoyed the most attention from scholars thus far.

The reader will notice that throughout this book we will make very few originality claims. This is partly because such claims are always tricky, and it seems that they can only be justified in case the philosophical context in which the author in question wrote, has already been thoroughly discussed in the scholarship. This is not the case for post-medieval modal logic, however. The fact that this is the first book-length study to appear on the topic says it all. The second reason is because in all probability Fabri was not the most original thinker. Fabri was a representative of a current – the *modernist* current, we shall argue – and while his work does contain a clear exposition of some of the ideas that were common within this current, he likely did not add many ideas of his own. This is, of course, not a reason *not* to devote a case study to his logic. If all research into the history of logic were to focus on big names like Aristotle, Ockham, Buridan, and Leibniz, then we will never be able to fully understand how the discipline evolved over time. A history of logic without any gaps (to echo the splendid book and podcast series by Peter Adamson) should, indeed, "avoid skipping from highlight to highlight", and also devote attention to the lesser gods.[17]

[17] Adamson (2014: xii).

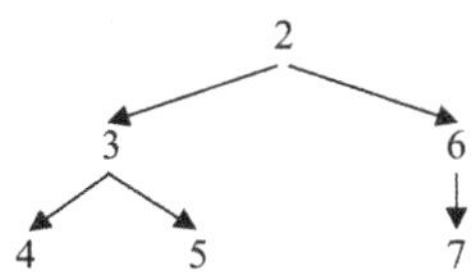

Fig. 1.1 Dependency relations between the different chapters

The main concern of this book is with Fabri's account of the syntax and truth conditions of, and the opposition and entailment relations between, modal propositions. These subjects are treated for composite modal propositions in Chaps. 3 and 4, and for divided modal propositions in Chap. 6. In both cases we also provide considerable historical background: Fabri's logic of composite modals is contextualized in Chap. 5, and his logic of divided modals in Chap. 7. These two chapters contain a wealth of data on how *antiqui* and *moderni* approached certain issues in modal logic. The reader who has a general interest in the history of modal logic yet does not necessarily want to know all details about a figure as peculiar as Fabri, might find these chapters particularly enjoyable. Chapter 2 is introductory in nature, and gives a reconstruction of the foundations of Fabri's modal logic. Chapter 8 wraps up the discussion, and draws some conclusions. Since Chaps. 2, 3 and 4 and Chap. 6 mainly draw on Fabri's questions on *De interpretatione*, it has seemed useful to also include a transcription of the relevant parts of this commentary, which is found in Appendix I. All references of the form '**S**, ll. *x-y*' in the footnotes are to this transcription. Figure 1.1 gives a visualization showing how the different chapters are related to each other. Notice that a complete overview of Fabri's modal logic would also require a study of his modal syllogistic, which is set out in the questions on the *Analytica priora* in **S**, at ff. 238va-244va. This topic will not be discussed here. Also, all translations in the book are our own, unless otherwise indicated.

Finally, a couple of words on methodology. Most studies in the history of logic adopt one of two methods, and approach their sources either *historically* or *rationally*.[18] When sources are approached historically, they are studied in relation to their broader philosophical context. The defining characteristic of historical reconstructions is that they only rely on concepts and tools that were available to the original author. Thus they are typically very nuanced and rich in philosophical content. By contrast, when sources are approached rationally, they are studied using tools of contemporary logic. These tools are often mathematical in nature, and thus rational reconstructions usually offer very precise results which can also be relevant for contemporary logicians. The main drawback of the rational approach is that we run the risk of anachronism: since highly mathematical ideas are 'projected onto' a historical system, we might attribute ideas to an author that are not entirely faithful to their thought. The approach taken in this book is predominantly historical – this is the first study to appear on Fabri's logic and philosophy of language, and thus it would make little sense to start formalizing his thought straight away. We will occasionally rely on contemporary (modal) logic, but with the

[18] See esp. Cameron (2011) for a discussion of these concepts.

exception of Chap. 4 the use of formal methods is minimal. The hurried reader who is not particularly interested in rational reconstruction can skip this chapter without any problem. When we engage in rational reconstruction, we make use of the following notation, which is an extension of that of Thom (2003).

'$\mapsto$'	'is included in'	'$\not\mapsto$'	'is not included in'
'$\smile$'	'overlaps with'	'$\mid$'	'excludes'
'$*$'	'necessarily', 'necessary' (governing a term)	'$\Box$'	'necessary' (governing a proposition)
'$\dagger$'	'possibly', 'possible' (governing a term)	'$\Diamond$'	'possible' (governing a proposition)
'$\ddagger$'	'contingently', 'contingent' (governing a term)	'∇'	'contingent' (governing a proposition)
'$\S$'	'impossibly', 'impossible' (governing a term)	'$\bigcirc$'	'impossible' (governing a proposition)
'$\sim$'	'not' (governing a term or a modality attaching to a term)	'$\neg$'	'not' (governing a proposition)

Thus '$^{\sim}*$' reads 'not necessarily', and '$^{\sim\dagger}$' reads 'not possibly'; while '$\neg\Box$' reads 'not necessary', and '$\neg\Diamond$' reads 'not possible'. The upper-case letters 'A', 'B' and 'C' are schematic variables for common terms (one-place predicates), and the lower-case letters 'a', 'b' and 'c' are schematic variables for singular terms (individual constants) resp. falling under 'A', 'B', 'C'. If a variable is underlined, then it has existential import: the formula in which it appears is true only if (the substituent of) that variable has at least one *suppositum*, i.e. is non-empty. The Greek letters 'φ', 'χ', and 'ψ', possibly with subscripted numbers added, are schematic variables for propositions (token-sentences). Round brackets around propositional variables, as in '(φ)' and '(χ)', are used to indicate the nominalizations of the propositions these variables represent. The symbol '$\rightarrow$' is the material implication, and '$\leftrightarrow$' is the material equivalence. The table below gives some examples of formulas and their informal semantics.

$\underline{A} \mapsto B^{\sim *}$	'Every A is not necessarily B' (only true if A is non-empty; alternatively: $\underline{A} \mid B^{*}$)	$\neg\Box(A \mapsto B)$	'That every A is B is not necessary'
$A \mapsto B^{*\sim}$	'Every A is necessarily not B'	$\Box(A \mid B)$	'That every A is not B is necessary'
$\underline{A} \smile B^{\ddagger}$	'(Some) A is contingently B' (only true if A is non-empty)	$\nabla(A \smile B)$	'That (some) A is B is contingent'
$A \not\mapsto B^{\S\sim}$	'(Some) A is not impossibly not B' (alternatively: $A \smile B^{\sim\S\sim}$)	$\neg\bigcirc(A \not\mapsto B)$	'That (some) A is not B is not impossible' (alternatively: $\neg\bigcirc(A \smile B^{\sim})$)

(continued)

$\underline{a} \not\twoheadrightarrow B^{\dagger\sim}$	'This *A* is not possibly not *B*' (only true if *a* is non-empty; alternatively: $\underline{a} \mid B^{\dagger\sim}$)	$\neg\Diamond(a \not\twoheadrightarrow B)$	'That this *A* is not *B* is not possible' (alternatively, $\sim\Diamond(a \mid B)$)
		$\neg\Diamond(\varphi)$	'That φ is not possible'
		$\Diamond(\neg\varphi)$	'That $\neg\varphi$ is possible'

Note that if a formula contains an adverbial modality, as is the case in the formulas on the left-hand side in the above table, then the golden rule is to interpret the modality first, and the binary operator placed between the schematic variables afterwards. For instance, in the case of '$A \not\twoheadrightarrow B^{\dagger}$', we should first interpret '$B^{\dagger}$', which reads 'possibly *B*', and then '$A \not\twoheadrightarrow B$', which reads 'Some *A* is not *B*'. We get the interpretation of '$A \not\twoheadrightarrow B^{\dagger}$' by putting the two together: 'Some *A* is not possibly *B*'. Likewise, in the case of '$A \smile B^{\dagger\sim}$', we first interpret '$B^{\dagger\sim}$' ('possibly not *B*'), and then we interpret '$A \smile B$' ('Some *A* is *B*'). Thus, '$A \smile B^{\dagger\sim}$' reads 'Some *A* is possibly not *B*'.

References

Adamson P (2014) A history of philosophy without any gaps: classical philosophy. Oxford University Press, Oxford

Ashworth EJ (2016) The post-medieval period. In: Novaes CD, Read S (eds) The Cambridge companion to medieval logic. Cambridge University Press, Cambridge, pp 166–191

Bartocci B (2016) In Albert's long shadow. Reading Aristotle's *Topics* in fifteenth-century schools. Rivista di Filosofia Neo-Scolastica 108:857–877

Bartocci B (2017) Dialectical reasoning and topical argument in the middle ages. An inquiry into the commentaries on Aristotle's *Topics* (1250–1500). Unpublished PhD Dissertation, University of Tours, Tours

Bartocci B, Masolini S (2014) Reading Aristotle at the University of Louvain in the fifteenth century. A first survey of Petrus de Rivo's commentaries on Aristotle (II). Bulletin de philosophie médiévale 56:281–383

Baudry L (ed) (1950) La Querelle des Futurs Contingents (Louvain 1465–1475). Textes inédits. Vrin, Paris

Biard J (2010) Nominalism in the later middle ages. In: Pasnau R (ed) The Cambridge history of medieval philosophy, vol 2. Cambridge University Press, Cambridge, pp 661–673

Cameron M (2011) Methods and methodologies. An introduction. In: Cameron M, Marenbon J (eds) Methods and methodologies. Aristotelian logic east and west, 500–1500. Brill, Leiden, pp 1–24

Ehrle F (1925) Der Sentenzenkommentar Peters von Candia, des Pisaner Papstes alexanders V. Ein Beitrag zur Scheidung der Schulen in der Scholastik des vierzehnten Jahrhunderts und zur Geschichte des Wegestreites. Aschendorffschen Verlagsbuchhandlung, Münster

Geudens C (2018) On topical logic during the late middle ages. A study of Saint-Omer, BA., Ms. 609 (c. 1502). Bulletin de philosophie médiévale 60:81–196

Geudens C (2020) Louvain theories of topical logic (c. 1450–1533). A reassessment of the traditionalist thesis. Unpublished PhD Dissertation, University of Louvain, Louvain

Guerlac R (trans) (1989) The Quarrel over Future Contingents (Louvain 1465–1475). Kluwer, Dordrecht

Hoenen MJFM (2003) *Via antiqua* and *via moderna* in the fifteenth century. Doctrinal, institutional, and church political factors in the *Wegestreit*. In: Friedman RL, Nielsen LO (eds) The medieval heritage in early modern metaphysics and modal theory, 1400–1700. Springer, Dordrecht, pp 9–36

Kałuża Z (1988) Les querelles doctrinales à Paris. Nominalistes et réalistes aux confins du XIVe et du XVe siècle. Pierluigi Lubrina, Bergamo

Kałuża Z (1998) Late medieval philosophy, 1350-1500. In: Marenbon J (ed) Routledge history of philosophy, vol. 3: Medieval philosophy. Routledge, London, pp 426–451

Kinney D (ed, trans) (1963) The complete works of St. Thomas More, vol. 15. Yale University Press, New Haven, CT

Madeira J (2006) Pedro da Fonseca's *Isagoge philosophica* and the Predicables from Boethius to the *Lovanienses*. Unpublished PhD dissertation, University of Leuven, Leuven

Marenbon J (2012) Latin philosophy, 1350-1550. In: Marenbon J (ed) The Oxford handbook of medieval philosophy. Oxford University Press, Oxford, pp 220–244

Meuthen E (1988) Kölner Universitätsgeschichte, vol. 1: Die alte Universität. Böhlau, Köln

Papy J (1999) The reception of Agricola's *De inventione dialectica* in the teaching of logic at the Louvain Faculty of Arts in the early sixteenth century. In: Akkerman F et al (eds) Northern humanism in European context, 1469–1625. From the 'Adwert Academy' to Ubbo Emmius. Brill, Leiden, pp 167–185

Reusens EHJ (1867) Statuts primitifs de la Faculté des Arts de Louvain. Compte rendu des séances de la Commission Royale d'Histoire 9:147–206

Ritter G (1921–1927) Studien zur Spätscholastik, 3 vols. Winter, Heidelberg

Schabel C (1995) Peter de Rivo and the Quarrel over Future Contingents at Louvain: new evidence and new perspectives. Part I. Documenti e studi sulla tradizione filosofica medievale 6:363–473

Schabel C (1996) Peter de Rivo and the Quarrel over Future Contingents at Louvain: new evidence and new perspectives. Part II. Documenti e studi sulla tradizione filosofica medievale 7:369–435

Schillings A (ed) (1962) Matricule de l'Université de Louvain, vol. 3: 31 août 1485–31 août 1527. Palais des Académies, Bruxelles

Swieżawski S, Prokopowicz M (1990) Histoire de la philosophie européenne au XVe siècle. Beauchesne, Paris

Thom P (2003) Medieval modal systems. Problems and concepts. Ashgate, Aldershot

Verbeke D (2013) Maarten Van Dorp (1485-1525) and the teaching of logic at the University of Leuven. Humanistica Lovaniensia 62:225–246

Chapter 2
The Foundations of Fabri's Modal Logic

Abstract This chapter introduces the basic elements of Fabri's modal logic. The main subjects that are addressed are the syntax of modal propositions and the extension of the set of modal operators. We point out that Fabri considers epistemic terms and propositional attitudes as modal operators, and after charting some parallels in contemporary sources we conclude that this is a first indication that Fabri's modal logic belongs to the modernist side of the *Wegestreit*. This chapter thus sets the stage of what follows. It introduces one of the main claims of the book, and it equips the reader with the background needed to understand the next chapters.

Keywords Modal logic · John Fabri of Valenciennes · *Wegestreit* · *Via moderna*

Fabri's modal logic is based on four modalities: the necessary (*necesse*; N), the possible (*possibile*; P), the impossible (*impossibile*; I), and the contingent (*contingens*; C). For ease of reference, it will be useful to collect these modalities into a set $\mathfrak{M}$:

$$\mathfrak{M} = \{N, P, I, C\}$$

That Fabri recognizes the members of $\mathfrak{M}$ as modalities is hardly surprising. They constitute the basis of Aristotle's modal logic, and they are included in virtually every medieval discussion of the topic. Yet does $\mathfrak{M}$ contain all modalities there are? The precise demarcation of the set of modalities had been a debated issue throughout the medieval period, and this was no different for the *Wegestreit*. Part of the trouble is due to Aristotle, who recognized the true (ἀληθές; *verum*) and the false (οὐκ ἀληθές; *falsum*) as modalities at *De int.* 12, 22a10-13. The remark is incidental, and the truth values are not actually used as operators at either *De int.* 12-13 or *An. pr.* I.3, 8-22, which are based on $\mathfrak{M}$ only. Yet due to Aristotle's authority many *Wegestreit* authors took it that whether the truth values have a modal character needs explaining. The matter was not made easier by the fact that someone like Peter of Spain (fl. c. 1220–1250) had incorporated Aristotle's remark in the first tract of his

C. Geudens, L. Demey, *The Modal Logic of John Fabri of Valenciennes (c. 1500)*,
SpringerBriefs in Philosophy, https://doi.org/10.1007/978-3-030-98802-9_2

Summulae logicales, a treatise that was vastly influential during the late medieval period.[1]

Moreover, all modalities in $\mathfrak{M}$ are alethic, and it was common knowledge among medieval authors that alethic terms behave similarly to epistemic terms such as 'to know' (*scire*), 'to believe' (*credere*), 'to imagine' (*imaginari*) and 'to opine' (*opinari*). This gave rise to a different debate, viz. whether propositional attitudes, too, have a modal character. The question was answered in the affirmative by Buridan and his associate Albert of Saxony (c. 1320–1390) around the mid-fourteenth century, and, earlier, by Ockham, who, in his *Summa logicae* and commentary on *De interpretatione*, had recognized as modalities all terms that can be predicated of an assertoric proposition, including propositional attitudes.[2]

The inclusion of epistemic terms among the modalities was a typical trait of modernist modal logics during the *Wegestreit*, occurring in authors such as Johann Eck (1486–1543), who was mainly active in Ingolstadt, and George of Brussels (†1510), Jerónimo Pardo (†1502), Juan Dolz (fl. c. 1510–1518), Fernando de Enzinas (†1523), Robert Caubraith (c. 1483–1544) and Juan de Celaya (c. 1490–1558), who all had ties with the circle of John Mair (1467–1550) at the Paris Collège de Montaigu, as well as in the anonymous *Compilatio ex Buridano, Dorp, Ockan et aliis nominalibus*, a commentary on Peter of Spain that was used in the Paris Collège de Navarre near the end of the fifteenth century.[3] The position was associated, with Ockham, by the Scotist Nicholas de Orbellis (†1475), and with Buridan, by Juan Luis Vives (1493–1540), who was an adept of humanist logic yet had studied at Montaigu.[4] In a 1469 Louvain commentary on the *De int.* 12-13, moreover, it is labelled as idiosyncratic of the *via moderna*.[5]

The subject of whether propositional attitudes qualify as modalities is not addressed by Fabri in his questions on *De interpretatione*, but propositional attitudes are treated as such in the questions on the *Analytica priora*. Occasioned by

---

[1] See Hispanus, *Summulae logicales* I.19 (de Rijk 1972: 11–12 [ll. 25, 1–6]); and Ferreira (1952) for a discussion of the aftermath of the *Summulae*.

[2] Buridanus, *in De int.* II, q. 7 (van der Lecq 1983: 77 [ll. 12–18); de Saxonia, *Perutilis logica* III.5 (Berger 2010: 448 [ll. 3–17]); Ockham, *in De int.* II.5 (Gambatese and Brown 1978: 461 [ll. 53–60]). There are some important differences between the Buridanian and the Ockhamist approach towards epistemic modalities, but these do not matter here. See Perini-Santos (2002) for further discussion.

[3] Bruxellensis, *in De int.* (Bruxellensis [1494?]: sig. m 3vb); Caubraith, *Quadrupertitum* (Caubraith 1516: f. 121va); [Compilatores], *in Sum. log.* ([Compilatores] [1495]: sig. d v^{va}); de Celaya, *in Sum. log.* (de Celaya 1525: M iii^{va}); Dolz, *in Sum. log.* (Dolz 1512: sig. S iii^{ra-b}); Eckius, *in Sum. log.* (Eckius 1516: f. 20rb); Enzinas, *in Sum. log.* (Enzinas 1528: f. 33rb); Pardo, *Medulla dyalectices* (Pardo 1505: f. 106rb). On Mair's circle, see esp. Broadie (1985); Noreña (1975: 1–35).

[4] De Orbellis, *in Sum. log.* (de Orbellis 1494: sig. c i^{rb}); Vives, *Adversus pseudodialecticos* (Guerlac 1979: 128).

[5] The Louvain commentary is preserved in Cambrai, Bibliothèque municipale, MS 962. The relevant passage occurs on f. 104^{v}: "Moderni tamen dicunt 'scitum', 'creditum', 'opinatum' esse modos etiam propositionis modalis propriissime capte ...". See Geudens (2020: 300–302) for a discussion of the commentary's authorship.

Aristotle's remarks on the conversion of modal propositions at *An. pr.* I.3, 25a27-25b26, Fabri asks whether and how conversion works for propositions where the copula is modified by the term '*scitum*', '*imaginatum*', or '*creditum*', evidently assuming that these terms are modal operators.[6] Further on in the commentary, while discussing the modal syllogistic, he also addresses the validity of syllogisms that involve one of these terms.[7] The epistemic modalities do not play a central role in Fabri's modal logic, however. They are not discussed outside these two passages, and for that reason they will not be taken into consideration in what follows. But the very fact that Fabri ascribes them a modal status is a first indication that his modal logic is of modernist inclination.

As for the truth values, Fabri denies that the true and the false are modalities, like most of his contemporaries. The reason has to do with the concepts of opposition and equipollence – '*equipollentia*' in Latin, a term that Fabri uses synonymously with '*equiualentia*' or equivalence. Propositions that are modified by the members of $\mathfrak{M}$ have specific opposition and equipollence relations holding between them, and propositions that are modified by the true or the false cannot be fitted into this scheme. So much is clear from quote [*a*] below. Fabri there replies to the objection that Aristotle's modal logic from *De int.* 12-13 is flawed on the grounds that the truth values are not treated as modal operators in these chapters, in spite of the remarks at *De int.* 12, 22a10-13.

> [*a*] It is argued: 'Some modal propositions are about truth and falsity ... But such propositions are not mentioned by Aristotle; and so Aristotle does not discuss the whole matter (*ergo ipse est diminutus*).' It is said that authors generally do not mention such propositions. This is because such propositions do not have equipollence or opposition relations with the modal propositions that are treated here. Only modal propositions involving the modes 'possible', 'contingent', 'impossible' and 'necessary' have equipollence and opposition relations holding between them, which are obtained by inserting negations in front of and after the modality (*per prepositionem et postpositionem negationum*).[8]

Fabri is unoriginal in invoking the concepts of equipollence and opposition to exclude the truth values from $\mathfrak{M}$. Similar arguments occur in many contemporary accounts of modal logic, by traditionalists and modernists alike.[9] Authors often added a further reason why the truth values should not be considered modalities. This further reason involves the concept of ampliation (*ampliatio*), or supposition widening. A commonly endorsed view among authors from the thirteenth century onwards held that the supposition of the subject term in a present-tense assertoric proposition is restricted to the present yet can be ampliated, or widened, by tinkering

[6] **S**, f. 231vb.

[7] **S**, f. $244^{ra\text{-}b}$.

[8] **S**, ll. 99–105.

[9] See, e.g., Bruxellensis, *in De int.* (Bruxellensis [1494?]: sig. m 3vb); Caubraith, *Quadrupertitum* (Caubraith 1516: f. 121va); Crockaert, *in Sum. log.* (Crockaert 1512: sig. d ii^{va}); de Monte, *in Sum. log.* (de Monte 1489: f. 30va); de Orbellis, *in Sum. log.* (de Orbellis 1494: sig. c $i^{ra\text{-}b}$); Soto, *in Sum. log.* (Soto 1529: f. 63ra); Dorp, *Perutile compendium* (Dorp 1499: sig. d 1ra); Tartaretus, *in Sum. log.* (Tartaretus 1498: f. 14vb); Versoris, *in Sum. log.* (Versoris 1572: f. 40^{v}).

with the proposition's copula, such that the term is made to stand for both present and past, future, possible or imaginary *supposita*.[10] Authors such as John Versoris († after 1482), Lambert of 's Heerenberg (†1499) and Caubraith justify their choice not to consider the truth values as modalities not only by referring to the principles of equipollence and opposition, but also by pointing out that while the members of $\mathfrak{M}$ can ampliate the supposition of the subject term to the possible, the truth values lack the ability to ampliate: they do not have the so-called *vis ampliandi*.[11] Fabri does not mention ampliation in this regard.

Like most of his contemporaries, Fabri interprets the modalities relative to time, and not relative to possible worlds, as is common in present-day modal logic.[12] The interpretation of the modalities, as well as their functioning in propositional contexts, depends on the kind of modal proposition in which they occur. Fabri structures his modal logic using the distinction, first introduced by Aristotle (at *SE* 166a22-32) and known in the Latin West as early as Abelard, between propositions that are modal in the composite sense (*in sensu composito*), or composite modals for short (*modales compositae*), and propositions that are modal in the divided sense (*in sensu diviso*), or divided modals for short (*modales divisae*). Each class of propositions involves a distinctive interpretation of the modalities, which is further discussed in Chapters 3, 4 and 6. The remainder of this chapter reconstructs Fabri's thoughts on the basic syntax of composite and divided modals. Since Fabri believes that modal propositions are construed from assertoric propositions, it is imperative to start with his account of the syntax of the latter class of propositions.

The study of Fabri's commentary on *De interpretatione* leaves no doubt that he took all well-formed assertoric propositions to be instances of one of the templates in scheme (*a*), which the reader will recognize from the categorical syllogistic.

$$(a) \qquad \left[\begin{array}{l}\text{Every}\\ \text{Some}\\ [\textit{Indefinite}]\\ \text{This}\end{array}\right. \quad A \quad \left[\begin{array}{l}\text{is}\\ \text{is not}\end{array}\right. \quad B$$

[10]On ampliation in general and its role in medieval theories of supposition, see, e.g., Kann (2016). For a rational reconstruction of the concept of ampliation, see Klima (1988: 85–110). On ampliation in late-medieval logic in particular, see Ashworth (1974: 89–92); Broadie (1985: 76–88).

[11]Caubraith, *Quadrupertitum* (Caubraith 1516: f. 121va); de Monte, *in Sum. log.* (de Monte 1489: ff. 29va, 30va); Versoris, *in Sum. log.* (Versoris 1572: f. 40^{v}). Note that Versoris is often referred to as "Versor" in the scholarly literature. There is archival evidence that his actual name was Versoris, however; see Geudens (2017).

[12]See esp. Binini (2022); Knuuttila (1993) for an overview of medieval approaches to modality.

Each of the four expressions from the leftmost list in (*a*) can be paired with each of the two expressions in the rightmost list, and so (*a*) contains eight different templates, which are listed below. Each template is assigned its own unique identifier, which appears in the left-most column.[13]

- (*A*) Every *A* is *B* (*Omne A est B*) Universal-affirmative
- (*E*) Every *A* is not *B* (*Omne A non est B*) Universal-negative
- (*I*) Some *A* is *B* (*Aliquod A est B*) Particular-affirmative
- (*O*) Some *A* is not *B* (*Aliquod A non est B*) Particular-negative
- ($^{+}$*INDF*) *A* is *B* (*A est B*) Indefinite-affirmative
- ($^{-}$*INDF*) *A* is not *B* (*A non est B*) Indefinite-negative
- ($^{+}$*S*) This *A* is *B* (*Hoc A est B*) Singular-affirmative
- ($^{-}$*S*) This *A* is not *B* (*Hoc A non est B*) Singular-negative

For ease of exposition, we assume that these templates together constitute the set $\mathrm{T_A}$.

$$\mathrm{T_A} = \{A, E, I, O, {}^{+}INDF, {}^{-}INDF, {}^{+}S, {}^{-}S\}$$

Note that in both his assertoric and modal logic Fabri treats indefinite propositions as equivalent to their corresponding particulars. Thus, under uniform substitution, instances of the *I*- and $^{+}$*INDF*-templates are equivalent to each other, as are instances of the *O*- and $^{-}$*INDF*-templates.

Fabri maintains that given an assertoric proposition φ, a modal proposition is obtained from φ by first defining the *dictum* – or the accusative-infinitive construction, expressible by a that-clause in English – that corresponds to φ; and by then either attaching a modality to the *dictum* or inserting a modality inside the *dictum*, in between the *dictum*'s subject term and copula. The type of modal proposition that is obtained by *attaching* a modality to φ's *dictum* is what Fabri calls a composite modal; while the type of modal proposition that is obtained by *inserting* a modality inside φ's *dictum* in the manner explained is what he calls a divided modal. Templates (1) and (2) below are examples of templates for resp. composite and divided modals that Fabri would accept. Both templates are obtained from the *A*-template in $\mathrm{T_A}$.

(*A*) Every *A* is *B* (*Omne A est B*) ⇒ (*A*; *dictum*) That every *A* is *B* (*Omne A esse B*)

[13] Note that the quantifiers in scheme (*a*) are defined up to equivalence. Thus, while Fabri uses 'not every' (*non omnis*) as equivalent with 'some not' (*aliquis non*), and 'no' (*nullus*) with 'every not' (*omnis non*), these combinations are not included in (*a*).

(1) That every *A* is *B* is possible (*Omne A esse B est possibile*)
(2) For every *A* it is possible to be *B* (*Omne A possibile est esse B*)

Divided modals, however, need not be construed on the basis of a *dictum*, Fabri believes. Such modals can also be directly obtained from assertoric propositions, by simply inserting a modality in front of the proposition's copula. That is, given an assertoric proposition φ and a modality m, we can construe *two* divided modal propositions – one with and one without a *dictum* – that are equivalent to each other. More generally, for every template for divided modals that contains a *dictum* and is syntactically well formed according to Fabri's standards, there is a corresponding template that is equally well formed yet does not contain a *dictum*; and under uniform substitution, instances of these two templates are equivalent to each other. The template that corresponds to (2) in this sense is (2′).

(2′) Every *A* is possibly *B* (*Omne A possibiliter est B*)

Fabri introduces the distinction between composite and divided modal propositions in the passage in quote [*b*].

> [*b*] First we should note that modal propositions are twofold. Some are called composite modals, and [i] they are those propositions in which the modality is one of the extremes, either the subject term or the predicate term. Others are divided modals, where the modality and the verb are placed in between the two extremes; and [ii] what precedes the modality and the verb is then the subject term, or behaves like the subject term (*est subiectum uel tenens se ex parte subiecti*), while what follows the modality and the verb is the predicate term, or behaves like the predicate term (*est predicatum uel tenens se ex parte predicati*). [iii] For instance, in 'For a man it is possible to be white' (*hominem possibile est esse album*), 'man' is the subject and 'white' is the predicate, and the aggregate 'possible to be' (*possibile est esse*) is the copula or behaves like the copula (*est copula uel tenens se ex parte copule*).[14]

Fabri develops the distinction between composite and divided modals using the grammatical categories of subject and predicate term. This practice was common throughout the medieval period, occurring already in Aquinas in the thirteenth century (although Aquinas talks about modal propositions *de dicto* and *de re*, and not about modal propositions in the composite and divided sense) and later also in Walter Burley (c. 1275–1344) and Buridan, among others.[15] Following clause [i], the modality and the *dictum* are the two extremes of a composite modal, and they can both occur as subject and predicate term. However, in virtually all composite modals occurring in the questions on *De int.* 12-13 and *An. pr.* I.3, 8-22, the modality is the predicate term and the *dictum* is the subject term, as in template (1) above. The main,

---

[14] S, ll. 4–11.

[15] Aquinas, *De propositionibus modalibus* (Bocheński 1940: 193); Buridanus, *in De int.* II, q. 7 (van der Lecq 1983: 77 [ll. 18–26]), *Tractatus de consequentiis* II.2 (Hubien 1976: 57); Burlaeus, *De puritate artis logicae tractatus brevior* (Boehner 1955: 239 [ll. 22–28]). See esp. van der Lecq (1981) for further details.

if not only, reason why Fabri recognizes the well-formedness of propositions where the modality is the subject term and the *dictum* is the predicate term is probably because unlike someone like Crockaert, for instance, he endorses the validity of the conversion procedure of composite modals where the *dictum* and the modality swap places.[16] If Fabri had endorsed modal conversion yet restricted the class of composite modals to those propositions where the *dictum* is the subject and the modality is the predicate, then he would have committed to the view that conversion can result in ill-formed propositions, which is clearly problematic.

The case of divided modals, which is discussed at [ii], is more complex. Fabri uses disjunctions to characterize the constituents of such propositions. This should be understood in connection with the fact that divided modals can be construed either with or without a *dictum*: the disjunct 'is' applies to those without a *dictum*, while the disjunct 'behaves like' applies to those with a *dictum*. If a divided modal does not contain a *dictum*, then the aggregate 'is (not) *m*-ly (not)' is the copula (with $m \in \mathfrak{M}$), and the term preceding (resp. following) the copula *is* the grammatical subject (resp. predicate) of the proposition. Moreover, notice that the term preceding (resp. following) the copula is also the logical subject (resp. predicate) of the proposition. So in divided modals without a *dictum* – as in those composite modals where the *dictum* is the subject and the modality the predicate, for that matter – the categories of grammatical and logical subject and predicate coincide. This is not the case with divided modals with a *dictum*. In such propositions, the *dictum* is the grammatical subject, the modality is the grammatical predicate, and the conjugated form of 'to be' is the grammatical copula. But the logical copula of the proposition is the aggregate 'is (not) *m* (not) to be' (with $m \in \mathfrak{M}$), just as the term preceding (resp. following) the logical copula is the logical subject (resp. predicate). These terms *behave as if* they are the grammatical subject, copula and predicate, although in reality they are not. The structure of templates (2) and (2′) is thus as follows. ('Gram.' is short for 'grammatical', and 'log.' is short for 'logical'.)

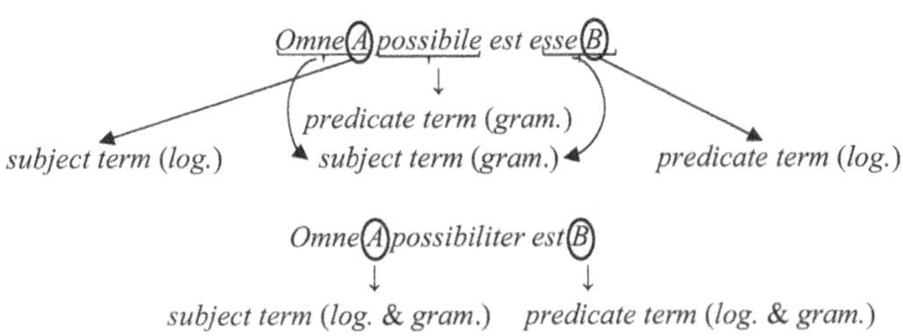

Thus the claim that 'man' and 'white' are the subject and predicate in the example at [iii] should be understood in terms of *logical* subject and predicate, as the

[16] Crockaert, *in Sum. log.* (Crockaert 1512: sig. d ii^{va}); **S**, ff. 231ra-232vb.

grammatical subject of the proposition is the *dictum* 'for a man to be white' (*hominem esse album*), and the grammatical predicate is the modality 'possible'. Further notice that by saying that the modality forms part of the logical copula of divided modals, Fabri is aligning himself with Mair, Crockaert, Nicolas of Amsterdam († before 1443), Antonio Coronel († c. 1520), and John Dullaert (1480–1513), as well as Buridan and his commentator John Dorp (fl. 1393–1418), *contra* Caubraith, Domingo de Soto (1494–1560), Peter Tartaret (c. 1460–1522) and the Vienna *modernus* Conrad Pschlacher (†1541), among others, who deny that the copula can be anything else than a form of 'to be' (*esse*), and thus maintain that a modality can only *determine* a copula, but is never a *part* of it.[17]

The main elements of the discussion up to this point are summarized in statements $\mathbf{A}_{1\text{–}3}$ below, which conclude this chapter. These statements interpret Fabri's modal logic as a system that deals with assertoric and modal propositions. Furthermore, the class of modal propositions consists of composite and divided modal propositions.

- $\mathbf{A}_1$: A proposition is *assertoric* iff it is an instance of one of the templates in T_A.
- $\mathbf{A}_2$: A proposition is *modal* iff it is either a composite modal proposition or a divided modal proposition.
- $\mathbf{A}_3$: Modalities include both *alethic* as well as *epistemic* terms.

References

Ashworth EJ (1974) Language and logic in the post-medieval period. Reidel, Dordrecht

Berger H (ed, trans) (2010) Albert von Sachsen: *Logik*. Felix Meiner, Hamburg

Binini I (2022) Possibility and necessity in the time of Peter Abelard. Brill, Leiden

Bochenski IM (1940) Thomas Aquinas: *De propositionibus modalibus*. Angelicum 17:180–218

Boehner P (ed) (1955) Walter Burleigh: *De puritate artis logicae tractatus longior*. With a revised edition of the *Tractatus brevior*. The Franciscan Institute, St. Bonaventure, NY

Bos EP (ed) (2016) Nicholas of Amsterdam: commentary on the *Old Logic*. John Benjamins, Amsterdam

Broadie A (1985) The circle of John Mair. Logic and logicians in pre-reformation Scotland. Clarendon Press, Oxford

Bruxellensis G [1494?] Cursus optimarum questionum super totam logicam cum interpretatione textus secundum viam modernorum ac secundum cursum magistri Georgii per magistrum Thomam Bricot sacre theologie professorem emendate. [s.n.], [s.l.]

Caubraith R (1516) Quadrupertitum in oppositiones, conuersiones, hypotheticas et modales. In aedibus Iodoci Badii et Emundi Fabri, Parrhisiis

[17] Buridanus, *in De int.* II, q. 6 (van der Lecq 1983: 74–76), *Summulae de dialectica* I.8.3 (van der Lecq 2005: 86–87 [ll. 30–31, 1–21]); Caubraith, *Quadrupertitum* (Caubraith 1516: f. 122vb); Coronel, *Prima pars rosarii* (Coronel 1517: sig. g i^{ra}); Crockaert, *in Sum. log.* (Crockaert 1512: sig. d ii^{va}); de Amsterdammis, *in De int.* (Bos 2016: 336–337 [ll. 30–31, 1–3]); Dorp, *Perutile compendium* (Dorp 1499: sig. d $2^{ra\text{-}b}$); Dullaert, *in De int.* (Dullaert 1515: f. 125va); Major, *Introductorium perutile* (Major 1527: f. 66ra); Pschlacher, *in Sum. log.* (Pschlacher 1516: f. 40^{r}); Soto, *in Sum. log.* (Soto 1529: f. 63vb-64ra); Tartaretus, *in De int.* (Tartaretus 1503: f. 55vb). For Fabri, see **S**, ll. 155–169. See also the discussion in Coombs (1990: 29–32).

[Compilatores] [1495] Compilatio ex Buridano, Dorp, Ockan et aliis nominalibus in textum Petri Hyspani edita in regali collegio Nauerre Parisius. [Nicolaus Wolff], [Lyon]
Coombs JS (1990) The truth and falsity of modal propositions in Renaissance nominalism. Unpublished PhD dissertation, University of Texas, Austin, TX
Coronel A (1517) Prima pars rosarii. Venalia apud Oliuerium Senant, [Paris]
Crockaert P (1512) Summularum artis dialectice vtilis admodum interpretatio super textum magistri Petri Hispani. Per Johannem Cleyn, Lugduni
De Celaya J (1525) Expositio in primum tractatum Summularum Petri Hispani. Venalia prostant in Clauso Brunello, Parrhisiis
De Monte L (1489) Textus omnium tractatuum Petri Hispani etiam sincathegreumatum [sic] et paruorum logicalium cum copulatis secundum doctrinam diui Thome Aquinatis iuxta processum magistrorum Colonie in Bursa Montis regentium. [Henricus Quentell], [Köln]
De Orbellis N (1494) Summule philosophie rationalis seu logica secundum doctrinam doctoris subtilis Scoti. Per Michaelem Furter, Basileae
De Rijk LM (ed) (1972) Peter of Spain (Petrus Hispanus Portugalensis): *Tractatus*, called afterwards *Summule logicales*. First critical edition from the manuscripts. Van Gorcum, Assen
Dolz J (1512) Disceptationes super primum tractatum Summularum. Opera ac caracteribus Johannis de la Roche, Parisius
Dorp J (1499) Perutile compendium totius logice Ioannis Buridani cum preclarissima expositione. Per Petrum Joannem de Quarengiis, Venetiis
Dullaert J (1515) Questiones super duos libros Peri hermenias Aristotelis. Impresse per Stephanum Baland, [Lyon]
Eckius J (1516) In Summulas Petri Hispani extemporaria et succincta sed succosa explanatio pro superioris Germaniae scholasticis. Ex officina Millerana, Augustae Vindelicorum
Enzinas F (1528) Primus tractatus Summularum. Apud Reginaldum Chauldiere, Parisiis
Ferreira J (1952) As *Súmulas logicais* de Pedro Hispano e os seus comentadores. Colectânea de Estudos 3:360–394
Gambatese A, Brown S (1978) Expositio in librum Perihermenias Aristotelis. In: Guillelmi de Ockham opera philosophica II. The Franciscan Institute, St Bonaventure, pp 345–504
Geudens C (2017) Versoris, Johannes. In: Sgarbi M (ed) Encyclopedia of Renaissance philosophy. Springer, Cham. https://doi.org/10.1007/978-3-319-02848-4_572-2
Geudens C (2020) Louvain theories of topical logic (c. 1450–1533). A reassessment of the traditionalist thesis. Unpublished PhD Dissertation, University of Louvain, Louvain
Guerlac R (trans) (1979) Juan Luis Vives: *Against the pseudodialecticians*. A humanist attack on medieval logic. Reidel, Dordrecht
Hubien H (ed) (1976) Iohannis Buridani Tractatus de consequentiis. Publications universitaires, Leuven
Kann C (2016) Supposition and properties of terms. In: Novaes CD, Read S (eds) The Cambridge companion to medieval logic. Cambridge University Press, Cambridge, pp 220–244
Klima G (1988) Ars artium. Essays in philosophical semantics, mediaeval and modern. Hungarian Academy of Sciences, Budapest
Knuuttila S (1993) Modalities in medieval philosophy. Routledge, New York
Major J (1527) Introductorium perutile in Aristotelicam dialecticen duos terminorum tractatus ac quinque libros summularum complectens. In aedibus Ioannis Parvi et Aegidii Gorimontii, [Paris]
Noreña CG (1975) Studies in Spanish Renaissance thought. Martinus Nijhoff, Den Haag
Pardo H (1505) Medulla dyalectices omnes ferme grauiores difficultates logicas acutissime dissoluens. Per Guillermum Anabat, Parisius
Perini-Santos E (2002) L'extension de la liste des modalités dans les commentaires du *Perihermeneias* et des *Sophistici Elenchi* de Guillaume d'Ockham. Vivarium 40:174–188
Pschlacher C (1516) Paruorum logicalium liber succincto epitomatis compendio continens perutiles argutissimi dialectici Petri Hispani tractatus priores sex et clarissimi philosophi Marsilii logices documenta cum utilissimis commentariis. Impensis Leonardi & Lucae Alantse, Viennae

Soto D (1529) Summule. In officina Joannis Junte, Burgis

Tartaretus P (1498) Expositio super Summulis Petri Hyspani. Expensis Jacobi Maillet, Lugduni

Tartaretus P (1503) Expositio super textu logices Aristotelis. Per Lazarum de Soardis, Venetiis

Van der Lecq R (1981) Buridan on modal propositions. In: Braakhuis HAG, Kneepkens CH, de Rijk LM (eds) English logic and semantics from the end of the twelfth century to the time of Ockham and Burleigh. Ingenium, Nijmegen, pp 427–442

Van der Lecq R (ed) (1983) Johannes Buridanus: *Questiones longe super librum Perihermeneias*. Ingenium, Nijmegen

Van der Lecq R (ed) (2005) Johannes Buridanus: *Summulae de propositionibus*. Brepols, Turnhout

Versoris J (1572) Petri Hispani Summulae logicales cum clarissima expositione. Apud Franciscum Sansovinum, Venetiis

Chapter 3
Fabri's Logic of Composite Modals

Abstract This chapter gives a historical reconstruction of Fabri's logic of composite modals. We first study the main outlines of his modal semantics, which is based on his theory of signification (*significatio*) and supposition (*suppositio*). Next we discuss his views of the propositional quantity of composite modals, and we end with an overview of his account of the truth conditions of such propositions.

Keywords Modal logic · Composite modals · Supposition · Propositional quantity · *Positio inesse* · Quantification invariance

In the average modern propositional modal logic, the simplest modal expressions are strings of the form '$\Diamond p$' and '$\Box p$'. The letter 'p' is usually understood as a variable expressing a state of affairs – or a 'proposition' in the present-day sense of this term (i.e., an eternal Platonic entity residing in a Fregean third empire). In standard Kripke semantics such variables are interpreted intensionally, as the set of possible worlds where the state they express obtains. The modal operators '$\Diamond$' and '$\Box$' are, as it is often said, 'proposition-forming operators on propositions'.[1] In Fabri's modal logic, *dicta* occurring in composite modals also function as variables of some kind. The main aim of this chapter is to explain what kind of variable the *dictum* is, and which role the *dictum* plays in Fabri's truth conditions of composite modals. Note that this chapter forms a unity with Chap. 4 and 5. In Chap. 4, we will propose a model-theoretic reconstruction of some of the ideas mentioned in Chap. 3. In Chap. 5, we will situate Fabri's logic for composite modals in its historical context, and we will inquire into his sources.

[1] See, e.g., Chellas (1980: 41); Hughes and Cresswell (1996: 14).

C. Geudens, L. Demey, *The Modal Logic of John Fabri of Valenciennes (c. 1500)*, SpringerBriefs in Philosophy, https://doi.org/10.1007/978-3-030-98802-9_3

3.1 Signification, Supposition and *Dicta*

The reader trained in modern logic might expect the claim that *dicta* occurring in composite modals are variables to mean something along the following lines. Such *dicta* are the result of nominalizing assertoric propositions, and every such proposition, being a token-sentence, expresses a state of affairs. The *dictum* that is obtained from an assertoric proposition φ is then simply a variable expressing the same state of affairs as the one that is expressed by φ. For instance, take the composite modal 'That every man is running is possible' (*Omnem hominem currere est possibile*), which is an instance of template (1) mentioned in Chap. 2. The *dictum* is the that-clause 'That every man is running', and it is obtained from the assertoric proposition 'Every man is running' (*Omnis homo currit*). The latter proposition expresses the state of affairs that every man is running, and, so the account would go, this same state is expressed by the proposition's corresponding *dictum*. Since *dicta* have the exact same meaning as their underlying propositions, it would follow that our example 'That every man is running is possible' can be restated as 'Possibly, every man is running' (*Possibiliter, omnis homo currit*). More generally, the account implies that every composite modal is equivalent to a proposition where the modality is an adverb, and a proposition-forming operator on propositions. For instance, template (1) is equivalent to (1′).

(1′) Possibly, every *A* is *B* (*Possibiliter, omne A est B*)

An account like this one indeed occurs in some late-medieval sources. Such an account, which is briefly mentioned by Coronel and more elaborately stated by Soto, among others, involves what was known as simple supposition.[2] According to Soto, simple supposition (*suppositio simplex*) is the mode of supposition where the propositional context makes a term stand for its significate (*significatum*), which is a collection of objects from the *res extra*, yet the term also connotes the concept (*conceptus*) through which its significate is known to the intellect.[3] For instance, 'man' has simple supposition in 'man is a species' due to its being subjected to a term of second intention (viz. 'species'), and it is made to stand for each and every entity that was, is, or will be a man, while also appealing to the concept of manhood. In composite modals, Soto says, the *dictum* has simple supposition, and it is made to pick out the significate of its underlying assertoric proposition. This significate belongs to the *res extra*, and it is either a *complexe significabile*, which is one of the precursors of the modern concept of state of affairs, or, as Soto prefers, the component objects of a *complexe significabile*. That the *dictum* is made to stand for

---

[2]Soto, *in Sum. log.* (Soto 1529: ff. 63vb-64ra); Coronel, *Prima pars rosarii* (Coronel 1517: sig. g ii^{ra}).

[3]Soto, *in Sum. log.* (Soto 1529: f. $18^{va\text{-}b}$). See Ashworth (1974: 84–88; 2013) for context and details on Soto's theory of (simple) supposition.

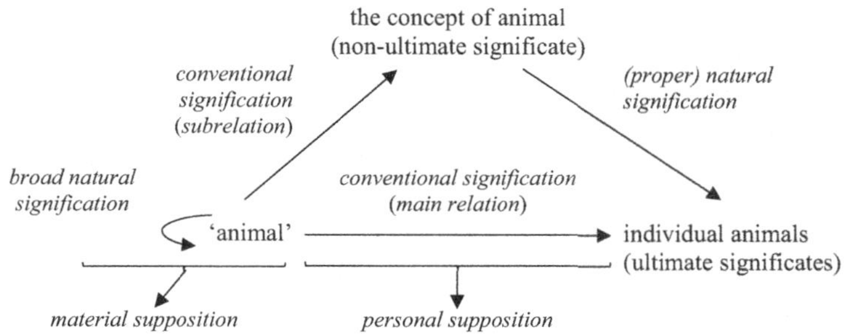

Fig. 3.1 Fabri's theory of signification, applied to the term 'animal'

the same significate as its underlying assertoric proposition implies, according to Soto, that each composite modal is equivalent to a proposition where the modality is an adverb. He gives the example of 'That every man is disputing is possible' (*omnem hominem disputare est possibile*), which he says is equivalent to 'Possibly, every man is disputing' (*possibiliter omnis homo disputat*).[4]

This is not the account that we find in Fabri, however. As opposed to Coronel and Soto, Fabri believes that *dicta* in composite modals supposit for linguistic items, and not for items from the *res extra*. By attaching a modal adjective to a *dictum*, the *dictum* is made to pick out the cluster of underlying assertoric propositions that actually exist at the moment of utterance, and not any kind of significate in the world out there.[5] For instance, in 'That every man is running is possible', if expressed at time *t*, the *dictum* 'That every man is running' supposits for all the propositions 'Every man is running' existing at *t*, and not for the state of affairs that every man is running or for the entities that are both men and running at *t*. In Fabri's semantics, this qualifies as a case of material supposition (*suppositio materialis*).[6]

Fabri's theory of supposition is embedded in his theory of signification. This latter theory is where we should look for an explanation of the notion that a *dictum*'s picking out its underlying assertoric propositions qualifies as a case of material supposition. Fabri nowhere gives a systematic treatment of the concept of signification, but we can roughly reconstruct his thoughts on the topic on the basis of his commentary on the opening chapters of *De interpretatione*. Every term *A*, assuming *A* has been subjected to an act of imposition, enters into two signification relations: *A* signifies conventionally (*significare ad placitum*), and *A* signifies naturally in a broad sense (*significare naturaliter communiter*) (see Fig. 3.1 for a visualization). The conventional signification relation terminates in the *res extra*, and is the composition of two subrelations: the first subrelation is the one where *A* signifies, also conventionally, its non-ultimate significate (*significatum non ultimatum*), and

[4] Soto, *in Sum. log.* (Soto 1529: f. 63[rb]).

[5] Medieval logics, it deserves bearing in mind, are generally token-based logics (see e.g. Kretzmann 1970). Fabri's (modal) logic is not an exception to this rule.

[6] See esp. **S**, ll. 130–143, 246–247 (see quote [*c*]).

the second subrelation is the one where *A*'s non-ultimate significate signifies, naturally and not conventionally, *A*'s ultimate significates (*significata ultimata*). These ultimate significates are items in the *res extra*. The non-ultimate significate is the concept that is formed from the ultimate significates by abstracting from sensory input.[7] Non-ultimate significates are objective concepts, Fabri emphasizes: 'They signify in the same way with everyone . . . such that if the concepts that reside in one man could be transferred to another man, then they would represent the same things for this other man.'[8] The broad natural signification relation terminates inside language, and not in the *res extra*. This relation is reflexive: *A* naturally signifies in a broad sense all occurrences of itself and of linguistic items similar to it. For instance, the term '*animal*' conventionally signifies the concept of animal (non-ultimate significate) as well as individual animals (ultimate significates), and it signifies naturally in a broad sense occurrences of '*animal*' and any of its inflexions, like '*animalis*' or '*animalibus*', among other things. Personal supposition (*suppositio personalis*), which is the mode of supposition where a term picks out entities from the *res extra*, is defined in terms of the conventional signification relation, and material supposition is defined in terms of the broad natural signification relation, as follows:

- *A* is made to supposit personally by a given propositional context just in case *A* is made to pick out, by this context, a subset of its ultimate conventional significates
- *A* is made to supposit materially by a given propositional context just in case *A* is made to pick out, by this context, a subset of its broad natural significates.[9]

Fabri's theory of signification is not original. Similar theories often occur in texts from the fifteenth and early sixteenth centuries, although in most accounts broad natural signification also involves the mediation of a concept, just like conventional signification. Fabri does not mention this concept, although it is hard to imagine that

[7] See **S**, f. 193rb: "... omnes uoces significatiue significant conceptus saltem tamquam significata non ultimata . . . ulterius significant sua principalia significata"; **S**, f. 193^{va-b}: "[Scripture, uoces et conceptus] habent se secundum ordinem, pro tanto quia conceptus magis uidentur ordinati ad significandum res, quia naturaliter significant et eodem modo apud omnes. Deinde uoces, quas licet oportet imponere ad placitum, tamen uidentur adhuc magis ordinate ad significandum res quam scripture; et ultimo scripture."

[8] See **S**, f. 194rb: "... conceptus sunt idem apud omnes in significando, puta in representando illa que naturaliter representant et significant, ut si conceptus qui sunt in uno homine possent transferri ad alium hominem, illi conceptus representarent illi homini idem"; along with **S**, f. 193va: "... conceptus que sunt in mente sunt idem apud omnes, supple in significando illa que naturaliter significant." Notice that, like many authors from the period, Fabri uses the terms 'to represent' (*repraesentare*) and 'to signify' (*significare*) interchangeably; see Ashworth (1990) for further discussion.

[9] See **S**, f. 193vb: "Notandum quod . . . quando [termini] supponunt pro seipsis uel †sibi similibus†, ut uoces pro scripturis uel conceptibus et econtra, talis suppositio uocatur materialis; quando uero supponunt pro significatis suis ultimatis, illa uocatur suppositio significatiua" (where '†...†' marks an uncertain reading); along with **S**, f. 195rb: "... ille uoces dicuntur significare naturaliter communiter que significant se uel similia in suppositione materiali, et sic omnes uoces mundi significant".

it does not form part of his theory. The characterization of material supposition in terms of broad natural signification goes back to the late fourteenth century, and may have been an innovation of Marsilius of Inghen and Pierre d'Ailly (1351–1420).[10]

So how do *dicta* fit in this theory? The conventional signification of a *dictum* collapses into the signification of the proposition from which the *dictum* is derived. A propositional significate is not a *complexe significabile*, Fabri maintains. Echoing a position that was taken by Ockham and Buridan, in the fourteenth century, and by Pardo and Mair, in his own day, Fabri endorses what Spade has dubbed the 'Additive Principle': the significate of a proposition is the sum total of the conventional significates of the proposition's categorematic constituent terms.[11] Thus, a proposition and its negation have the same conventional significate. Fabri gives the example of the propositions 'God is a being' (*Deus est*) and 'God is not a being' (*Deus non est*). These two propositions have the same conventional significate, viz. God along with all beings; and this significate is also the conventional significate of the *dicta* 'that God is a being' (*Deum esse*) and 'that God is not a being' (*Deum non esse*).[12] The broad natural significates of *dicta* are occurrences of themselves along with their underlying assertoric propositions. Thus, while two *dicta* deriving from a proposition and its negation have the same conventional significates, their broad natural significates differ. Some propositional contexts will make a *dictum* supposit materially for those broad natural significates that are occurrences of itself. Perhaps the most straightforward such context is the one where a *dictum* is subjected to the term '*dictum*', as in 'That every man is running is a *dictum*' (*omnem hominem currere est dictum*). By contrast, in composite modals, the propositional context makes the *dictum* supposit materially for all and only broad natural significates that are its underlying assertoric propositions.

It turns out that the imaginary modern reader from a couple paragraphs ago was quite wrong about Fabri. *Dicta* in composite modals actually have hardly anything in common with the propositional variables of today's average symbolic modal logic. Such variables express states of affairs, while *dicta* do not express structures in the *res extra* at all. In Fabri there is no trace of the idea that composite modals can be restated as propositions with a proposition-forming operator acting upon a proposition. If there is any element from modern logic which Fabri's *dictum* resembles, it is

[10] See Ashworth (1974: 83–84); Karger (1982); and, especially, Read (1999) for further details.

[11] Spade proposed the Additive Principle in a study on Ockham; see Spade (1975: 58–59). On Buridan, see e.g. Klima (2009: 203–209). On Mair and Pardo, see Ashworth (1978: 101) (reprinted in Ashworth 1985); Pérez-Ilzarbe (2004: 167–171).

[12] See **S**, f. 201va: "Replicatur: 'si aliqua enuntiatio esset falsa, maxime esset ista: 'Deus non est'; sed ista non est falsa, quod probatur quia quamcumque rem significat ista 'Deus est', illam significat ista 'Deus non est' et econtra; sed una est uera, ergo <et> altera'. Dicitur pro nunc (omissis aliis modis dicendi) quod licet quicquid significat una hoc significat altera, una tamen significat aliqualiter illam rem que est deus qualiter non significat altera, et illa alietas est in ipsa enunciatione et maxime in enuncatione mentali." (where '<...>' marks an insertion.) Formally, if $s(x)$ is the conventional significate of x, then $s(\mathit{Deus\ est}) = s(\mathit{Deus\ non\ est}) = \{s(\mathit{Deus}), s(\mathit{est})\} = s(\mathit{Deum\ esse}) = s(\mathit{Deum\ non\ esse})$.

a predicate variable. Like predicate variables, *dicta* can denote sets of entities, and their extension varies across circumstances (or is 'non-rigid'), although, of course, modern logic's predicate variables denote subsets of some assumed domain of quantification, i.e., sets of individuals, rather than sets of linguistic structures. To be more precise, Fabri's *dicta* are a kind of *metalinguistic* predicates, and the modal adjectives in composite modals, too, properly belong to the metalanguage rather than the object language.

3.2 The Quantity of Composite Modals

As we have seen, *dicta* in composite modals are complex (i.e., non-primitive) categoremes, resulting from the nominalization of assertoric propositions. While some late-medieval authors, like Soto, rely on (simple) supposition to have such *dicta* behave *as if* they are propositions, Fabri uses (material) supposition to have them behave in accordance with their grammatical nature as categoremes. One of the consequences of this view, he maintains, is that such *dicta* bear quantification.[13] That is, it is possible that the *dictum* d in a composite modal φ has a quantifier prefixed to it. This quantifier then indicates about how many of the propositions in the cluster for which d supposits – all of these propositions, some of them, or just one – an assertion is made in φ. Thus expressions such as (3), (4) and (5) below are not anomalous, but are among the templates for composite modals that Fabri would endorse, just as (1) already mentioned.

(1) That every A is B is possible (*Omne A esse B est possibile*)
(3) Every that every A is B is possible (*Omne omne A esse B est possibile*)
(4) Some that every A is B is possible (*Aliquod omne A esse B est possibile*)
(5) This that every A is B is possible (*Hoc omne A esse B est possibile*)

Clearly, any composite modal φ with a quantified *dictum* carries a different information content with respect to its corresponding composite modal ψ that has an unquantified *dictum*, or a *dictum* that is preceded by a different quantifier. But whether this difference in information content between φ and ψ comes with implications for the truth conditions of φ and ψ is an entirely different question, and we shall see in Sect. 3.3 that it should be answered in the negative. The procedure of quantifying over *dicta* is made possible by the supposition rules governing composite modals, but its added value will turn out to be rather limited.

[13] See esp. **S**, ll. 242–250 (see quote [*c*]).

$$(b)\quad \left[\begin{array}{l}\text{Every}\\ \text{Some}\\ [\textit{Indefinite}]\\ \text{This}\end{array}\right.\quad \text{that}\quad \left[\begin{array}{l}\text{every}\\ \text{some}\\ [\textit{indefinite}]\\ \text{this}\end{array}\right.\quad A\quad \left[\begin{array}{l}\text{is}\\ \text{is not}\end{array}\right.\quad B\quad \left[\begin{array}{l}\text{is}\\ \text{is not}\end{array}\right.\quad \left[\begin{array}{l}\text{necessary}\\ \text{possible}\\ \text{contingent}\\ \text{impossible}\end{array}\right.$$

Scheme (*b*) gives an overview of the syntactic templates on which Fabri's logic for composite modals is built. They are 256 in total ($4 \times 4 \times 2 \times 2 \times 4$). We assume that the templates in (*b*) together form the set T_{CM}, which is the analogue for composite modals of the set T_A. The attentive reader will notice that (*b*) is an extension of scheme (*a*) given earlier in the sense that scheme (*a*) occurs embedded in scheme (*b*).[14] Note, however, that the templates in T_{CM} constitute only a subset of the totality of templates that Fabri would accept as well formed. As we have seen in Chap. 2, Fabri also assigns a modal status to some epistemic terms, meaning the right-most list in (*b*) is incomplete, and he also accepts the well-formedness of some of the converses of the templates in T_{CM}. For the sake of clarity it should also be noted that a quantifier attaching to a *dictum* behaves rather differently from a quantifier attaching to the subject term inside a *dictum*. In any instance of one of the templates in T_{CM} that contains multiple quantification, the subject term *A* of the *dictum* has personal supposition, and thus picks out (some of) the entities from the *res extra* that presently fall under it. Thus, the quantifier attaching to the subject term ranges over non-linguistic entities, and says how many of the *supposita* of *A* – some of them, all of them, or just one – are among the (personal) *supposita* of the *dictum*'s predicate term *B*. By contrast, the *dictum* itself has material supposition, and is thus made to pick out (some of) its underlying assertoric propositions (cf. Sect. 3.1). Consequently, the quantifier attaching to the *dictum* ranges over these assertoric propositions, and says how many of them are among the propositions for which the modality *m* supposits. The *supposita* of *m*, moreover, are personal *supposita*, not material *supposita*. Fabri never discusses the supposition of *m* in the passages that we studied, but since these *supposita* constitute a subset of the entities that are determined by the non-ultimate significate of *m*, they should qualify as personal *supposita* by Fabri's theory of supposition. True, the personal *supposita* of *m* are linguistic structures, and in that sense they are rather different from the personal *supposita* of categoremes like 'man' and 'animal'. This, however, is precisely what we expect given that *m* is a predicate belonging to the metalanguage, and not to the object language. Notice, further, that *m* does *not* pick out occurrences of itself or of linguistic structures resembling itself, which would however be required were *m* to have material supposition.

It is obvious that on Fabri's account, composite modals are rather similar to assertoric propositions, as each template in T_A and T_{CM} reduces to a sequence of a

[14] Note that the quantifiers in (*b*) are defined up to logical equivalence, like the ones in (*a*). Thus, while Fabri occasionally writes composite modals of the form 'No that every *A* is *B* is possible', for instance, this template is not included in (*b*) since it is equivalent to 'Every that every *A* is *B* is not possible', which does occur in (*b*).

subject term – which may or may not be preceded by a quantifier –; a copula – which may or may not be preceded by a negation sign –; and a predicate term. That is, both composite modals and assertoric propositions possess the subject-copula-predicate structure. True, one of the extremes in a composite modal is always a complex categoreme, which is not true of the subject term in an assertoric proposition; and while none of the extremes in an assertoric proposition belong to the so-called logical vocabulary, one of the extremes in a composite modal – the modality – does belong to this vocabulary. But these differences aside, composite modals and assertoric propositions share the same syntactic structure. Specifically, every proposition that can be parsed according to a template in T_{CM}, i.e. every composite modal, can also be parsed according to a template in T_A, i.e. as an assertoric proposition. Fabri takes this syntactic similarity as proof that composite modals are not 'genuine' modal propositions, but are rather assertoric propositions in disguise. So much is clear from the passage in quote [*c*] below. In this passage, which is the main textual ground for the account sketched in this section, Fabri discusses the relation between composite modals and assertoric propositions, as well as the functioning of *dicta* in composite modals and the quantity of such propositions.

> [*c*] It is doubted, first, whether composite modals are truly modal propositions (*utrum modales composite sunt uere modales*); and second, how we should understand the quantity of composite modals ... To the first doubt it is said that composite modals are not truly modal propositions (*propositiones modales composite non sunt uere modales*). This is proven, for those propositions in which the copula does not occur modified, are not modal propositions; and thus composite modals are not truly modal propositions. The minor premise is proven; for in the proposition 'That something white is_1 black is_2 impossible' (*album esse nigrum est impossibile*), for instance, the word 'is_2', which is the copula, does not occur modified. To the second doubt it is said that the quantity of composite modals is properly understood in the same way as the quantity of assertoric propositions (*quantitas attenditur proprie in modalibus compositis sicut in propositionibus de inesse*). This is proven, for composite modals are truly assertoric propositions, as has been said already. From this it is clear that a composite modal is properly universal if its subject term is a common term that is determined by a universal sign, as in 'Every that a man is running is possible' (*omne hominem currere est possibile*). Here the *dictum* 'that a man is running' (*hominem currere*) is a common term that supposits materially and is determined by a universal sign. A composite modal is indefinite if its subject term is a *dictum* occurring without a sign, as in 'That a man is running is possible' (*hominem currere est possibile*); and it is singular if the subject term is a *dictum* with a demonstrative pronoun that indicates a primitive species (*primitiva species*), as in 'This that a man is running is possible' (*hoc hominem currere est possibile*).[15]

[15] S, ll. 234–250. The term '*species primitiva*' seems to be mostly used in the ancient and medieval grammatical tradition, where it indicates non-derivative terms (as opposed to so-called *species derivativa*). In quote [*c*], Fabri uses the term to refer to categorical propositions. Calling a proposition a *species primitiva* is a stretch, as propositions are complex linguistic entities, unlike terms. Fabri's choice of words might have to do with the fact that categorical propositions are primitive if compared to their *dicta*. On the concept of *species primitiva*, see Bursil-Hall (1971); Ebbesen (2019).

3.3 The Truth Conditions of Composite Modals

We can now address Fabri's account of the truth conditions of composite modals. Since Fabri considers composite modals as assertoric propositions, he holds that the truth or falsity of composite modals depends on the same principle as the truth or falsity of assertoric propositions, viz. co-supposition. The truth value of a composite modal φ at time t is thus a matter of whether the material *supposita* of φ's *dictum* at t are among the personal *supposita* of φ's modality at t. The nature of the material *supposita* of the *dictum* has been dealt with in Sect. 3.2, but the nature of the personal *supposita* of the modalities in $\mathfrak{M}$ has not yet been addressed. Thus, we should discuss this first, before we can turn to an examination of Fabri's truth conditions of composite modals. Fabri remains silent on the supposition of the modalities, as indicated, but we shall see that we can quite easily reconstruct his account based on his views on signification and supposition. After this issue has been clarified, we shall make two comments, one on existential import, and one on the use of quantifiers in composite modals. Then, finally, we will be in a position to address Fabri's thoughts on the truth conditions of these propositions.

For every modality $m \in \mathfrak{M}$, the set of personal *supposita* of m is a subset of the ultimate significates of m. These significates, in turn, are demarcated by m's non-ultimate significate. Remember, Fabri works with a temporal interpretation of modality, and thus the non-ultimate significate of m is a concept that involves a temporal element. Fabri does not treat the non-ultimate signification of the modalities in detail. There is one occasion where he interprets necessity as truth at all times (*semper uera*), indicating that the non-ultimate significate of 'necessary' is the concept of sempiternal truth.[16] This is the only passage in his questions on *De int.* 12-13 and *An. pr.* I.3, 8-22 that sheds light on the non-ultimate signification of the modalities; but it is enough to reconstruct what he takes to be the non-ultimate significate of the other modalities in $\mathfrak{M}$. On several occasions, Fabri makes a distinction between universal and particular modalities, using the labels 'universal' and 'particular' to indicate that the modalities so characterized distribute (universal), or do not distribute (particular), over times.[17] Fabri calls both the necessary and the impossible universal modalities, and thus, since 'necessary' signifies sempiternal truth, 'impossible' should signify sempiternal falsity (i.e., truth at no time). The possible is a particular modality, and thus 'possible' should signify truth at some time. Notice that the non-ultimate significates of the modalities all involve the notions of truth or falsity. The truth values will thus also occur in the conditions demarcating the personal *supposita* of each of these modalities. Remember, Fabri's

[16] **S**, ll. 297–307. Fabri's account thus seems to show similarities to the so-called 'statistical' or 'temporal frequency' interpretation of modality, which was developed in modern philosophical logic by Becker (1952) and applied to medieval discussions by Knuuttila (1993) *in primis*. For a critical discussion of the suitability of the statistical model to capture medieval concepts of modality, see Binini (2022).

[17] See esp. **S**, ll. 18–23, 27–35 (see quote [*p*]).

logic is a token-based system (see fn. 5 on p. 23); and that means, in particular, that a claim about the truth value of a proposition is always conditional upon the existence of that proposition. If we bear this in mind, then the following is an adequate characterization of the personal *supposita* of 'necessary', 'possible', and 'impossible'.

- The personal *supposita* of 'possible' at t are those propositions ψ existing at t, such that, for some t', if ψ or another occurrence of the same type as ψ exists at t', then it is true at t'.
- The personal *supposita* of 'necessary' at t are those propositions ψ existing at t, such that, for every t', if ψ or another occurrence of the same type as ψ exists at t', then it is true at t'.
- The personal *supposita* of 'impossible' at t are those propositions ψ existing at t, such that, for no t', if ψ or another occurrence of the same type as ψ exists at t', then it is true at t'.

The case of contingency is more complicated. Fabri adopts the two familiar readings of '*contingens*', one where it means the one-sided possible, and is thus equivalent with '*possibile*', and one where it means the two-sided possible, and is thus equivalent with the conjunction of '*possibile*' and '*non necesse*'.[18] If Fabri indeed takes the non-ultimate significates of 'possible' and 'necessary' to be truth at some time and sempiternal truth, respectively, then he takes the non-ultimate significate of 'contingent' to be truth at some time on the one-sided reading, and truth at some, but not all, times on the two-sided reading. If so, then he would agree that in composite modals the personal *supposita* of 'contingent' at t on the one-sided reading are the same as the personal *supposita* of 'possible' at t. He would also agree that the following is an adequate characterization of the personal *supposita* of 'contingent' on the two-sided reading, where this term 'supposits for all those propositions that are not necessary and not impossible' (*supponit pro omnibus propositionibus que non sunt necessarie neque impossibiles*).[19]

- The personal *supposita* of 'contingent' at t are those assertoric propositions ψ existing at t such that, for some t', if ψ or another occurrence of the same type as ψ exists at t', then it is true at t', and for some t'', if ψ or another occurrence of the same type as ψ exists at t'', then it is not true at t''.[20]

With the supposition of the modalities we have clarified one important element. But, as indicated, we should make two further comments before we can move on and

---

[18] See, e.g., **S**, ll. 209–214, 372–392.

[19] **S**, f. 241ra.

[20] The supposition of the modalities if used as predicate in categorical propositions where the subject is a simple term, like 'The Antichrist is possible' or 'All human beings are contingent', can be easily obtained from the above characterizations by substituting 'assertoric proposition' with 'being'. The nature of such propositions is a recurring theme in discussions on modal logic from the twelfth century onwards. Fabri briefly touches on the issue at **S**, ll. 393–399 but he does not discuss it in debt.

formulate Fabri's truth conditions of composite modals. The first one is that these truth conditions should include an existential import condition. For Fabri, affirmative assertoric propositions are false if their subject term is empty, while negative assertoric propositions are true in that case.[21] Composite modals classify as assertoric propositions, and so whether the *dictum* in a composite modal φ has at least one *suppositum* at t has implications for φ's truth value at t:

- If φ is affirmative, then φ is true at t only if there is at least one *suppositum* of φ's *dictum* at t
- If φ is negative, then φ is true at t if there is no *suppositum* of φ's *dictum* at t

where φ is affirmative just in case it involves a non-negated occurrence of 'necessary', 'possible', or 'contingent', or a negated occurrence of 'impossible', and φ is negative otherwise.[22] There is plenty of textual evidence in support of this interpretation. A telling passage is quote [*d*] below. There Fabri discusses arguments involving composite modals where the antecedent is negative, and the consequent is affirmative. Fabri points out that such arguments preserve truth only if there is at least one *suppositum* of the consequent's *dictum*, evidently assuming that the emptiness of the *dictum* would suffice for the consequent's falsity.

> [*d*] When we argue from negative propositions to affirmative ones, then the existence should be assumed of a proposition for which the *dictum* of the consequent supposits. For instance, we should argue as follows: 'that a man is a donkey is not possible' (*hominem esse asinum non est possibile*), and a proposition 'every man is not a donkey' (*omnis homo non est asinus*) exists; therefore, 'that every man is not a donkey is necessary' (*omnem hominem non esse asinum est necessarium*).[23]

As for the second comment, we first need a definition. If φ is a composite modal, then a *quantifier mate* of φ is a proposition that differs from φ in terms of propositional quantity, but is otherwise identical with φ. Thus, instances that are obtained

---

[21] See, e.g., **S**, f. 199va, where Fabri remarks, in the context of infinitized verbs like 'to not read' (*non legere*) and 'to not walk' (*non ambulare*), that the argument '*Iohannes non legit; ergo Iohannes non legit*' is invalid if the negation is a *negatio negans* in the antecedent ('it is not the case that John is reading') and a *negatio infinitans* in the consequent ('John is not-reading'), since 'the antecedent is true and the consequent false when John does not exist' (*si Iohannes non esset, antecedens esset uerum et consequens falsum*). Elsewhere, at **S**, f. 197vb, Fabri says that both 'Chimera is a man' (*Chimera est homo*) and 'Chimera is a not-man' (*Chimera est non homo*), where the negation in the second proposition is infinitizing, are false due to the non-existence of chimeras.

[22] Fabri discusses the quality of composite modals at **S**, ll. 250–254. He there says that 'those composite modals are affirmative in which the predicate term is affirmed of the subject term', while 'those composite modals are negative in which the predicate term is negated of the subject term'. This might suggest that composite modals with negated occurrences of 'impossible' class as negative. This is not the case, however. Elsewhere Fabri says that 'impossible' is a negative modality (*modus negatiuus*), and that propositions with non-negated occurrences of 'impossible' contain an 'implicit negation' (*implicite ponitur negatio*); see **S**, ll. 199–204. Thus composite modals with negated occurrences of 'impossible' are affirmative.

[23] **S**, ll. 292–295. There are several parallel passages in the texts where Fabri discusses modal conversion; see esp. **S**, f. $231^{va\text{-}b}$.

from templates (1), (3), (4) and (5) under uniform substitution are all quantifier mates of each other. We will show that in Fabri's modal logic, a composite modal is always equivalent to any of its quantifier mates with which it co-exists. This means, in particular, that we may neglect issues of quantification when we formulate truth conditions for composite modals. This equivalence of composite modals with their quantifier mates might appear counterintuitive, but it is in fact a straightforward consequence of Fabri's belief that truth depends on co-supposition and his account of the supposition of *dicta* in composite modals, as we shall now show. We will prove, first, that *affirmative* composite modals are equivalent with their quantifier mates, and, second, that *negative* composite modals are also equivalent with their quantifier mates. Since each composite modal is either affirmative or negative, we will thus have shown that *every* composite modal is equivalent with its quantifier mates.

For a start, let us focus on affirmative composite modals with non-negated modalities (i.e., propositions with non-negated occurrences of 'necessary', 'possible', or 'contingent'). Let φ be such a composite modal with universal quantity, let ψ be a quantifier mate of φ with particular quantity, and let χ be a quantifier mate of φ (and ψ) with singular quantity. Remember, composite modals are assertoric propositions. Thus, the truth conditions of φ, ψ and χ should not only be formulated in terms of co-supposition and include an existential import condition, as pointed out, but they should also interpret the quantifiers in φ, ψ and χ in the way quantifiers are ordinarily interpreted in assertoric logic. Consequently, the truth conditions of φ, ψ and χ are as follows:

- φ is true at t iff there is at least one *suppositum* of φ's *dictum* at t and all *supposita* of φ's *dictum* are among the *supposita* of φ's modality at t (i.e., conjunctive descent)
- ψ is true at t iff there is at least one *suppositum* of ψ's *dictum* at t and at least one *suppositum* of ψ's *dictum* is among the *supposita* of ψ's modality at t (i.e., disjunctive descent)
- χ is true at t iff the indexical picks out a *suppositum* of χ's *dictum* at t and that *suppositum* is among the *supposita* of χ's modality at t

As we have seen, Fabri restricts the set of *supposita* of the *dictum* in a composite modal to occurrences of its underlying assertoric proposition, and thus to occurrences of one and the same proposition type (see Sect. 3.1). This means that, as soon as we have clarity about the relation of *one* such *suppositum* with the *supposita* of the proposition's modality, then in fact we have clarity about the relation of *all* such *supposita* with the *supposita* of the modality: one *suppositum* of the *dictum* is among the *supposita* of the modality just in case all *supposita* of the *dictum* are.[24] This

[24]To make this a bit more concrete, consider a sect whose members all share exactly the same beliefs. Then we observe the same kind of collapse as between the *supposita* of the *dictum*: one member of the sect will have a certain belief iff all members of the sect have that belief.

implies that if φ, ψ and χ co-exist, then they constitute one equivalence class. Here is the proof.

(1) χ and φ are equivalent: Consider any time t and assume that χ and φ exist at t. ($\Rightarrow$) Assume that χ is true at t. Then, by the truth condition of χ and the fact that χ and φ are quantifier mates, there is a *suppositum* of φ's *dictum* that is among the *supposita* of φ's modality at t. Thus the *supposita* of φ's *dictum* are a non-empty subset of the *supposita* of φ's modality at t, by Fabri's restrictions on the supposition of the *dictum*, and so φ is also true at t. ($\Leftarrow$) Assume that φ is true at t. Then, by the truth condition of φ and the fact that φ and χ are quantifier mates, the *supposita* of χ's *dictum* are a non-empty subset of the *supposita* of χ's modality at t. Thus the reference of the indexical in χ is satisfied at t, and its referent is among the *supposita* of χ's modality at t. So χ is true at t. ∎

(2) ψ and φ are equivalent: Consider any time t and assume that φ and ψ exist at t. ($\Rightarrow$) Assume that ψ is true at t. Then, by the truth condition of ψ and the fact that φ and ψ are quantifier mates, there is at least one *suppositum* of φ's *dictum* that is among the *supposita* of φ's modality. Thus the *supposita* of φ's *dictum* are a non-empty subset of the *supposita* of φ's modality, by Fabri's restrictions on the supposition of the *dictum*, and φ is true at t. ($\Leftarrow$) Assume that φ is true at t. Then, by the truth condition of φ and the fact that φ and ψ are quantifier mates, the *supposita* of ψ's *dictum* are a non-empty subset of the *supposita* of ψ's modality, and so there is at least one *suppositum* of ψ's *dictum* that is among the *supposita* of ψ's modality. So ψ is true at t. ∎

By the transitivity of equivalence, ψ and χ are also equivalent, which finishes the proof that φ, ψ and χ constitute one equivalence class if they co-exist. Add to this that Fabri treats indefinite propositions as equivalent to particulars, and we get that *every* affirmative composite modal with a non-negated modality is equivalent with *all* its co-existing quantifier mates. The same holds for affirmative composite modals with negated modalities (i.e., with an occurrence of 'not impossible'), as the reader can check, and thus every affirmative composite modal *tout court* constitutes an equivalence class with its co-existing quantifier mates.

The proof for negative composite modals is analogous. Let us begin by focusing on negative composite modals with negated modalities (i.e., propositions with negated occurrences of 'necessary', 'possible', or 'contingent'). Let φ be such a composite modal with universal quantity, let ψ be a quantifier mate of φ with particular quantity, and let χ be a quantifier mate of φ (and ψ) with singular quantity. These propositions have the following truth conditions:

- φ is true at t iff there is no *suppositum* of φ's *dictum* at t or no *suppositum* of φ's *dictum* is among the *supposita* of φ's modality at t
- ψ is true at t iff there is no *suppositum* of ψ's *dictum* at t or at least one *suppositum* of ψ's *dictum* is not among the *supposita* of ψ's modality at t

- χ is true at t iff the indexical does not pick out a *suppositum* of χ's *dictum* at t or the *suppositum* picked out by the indexical is not among the *supposita* of χ's modality at t

By appealing to these truth conditions and to Fabri's restrictions on the supposition of the *dictum* in a composite modal, we can once again prove that φ, ψ and χ constitute one equivalence class if they co-exist. Furthermore, remember that Fabri treats indefinites as equivalent with particulars. Thus, *every* negative composite modal with a negated modality is equivalent with *all* its co-existing quantifier mates. The result can be extended to negative composite modals with non-negated modalities (i.e., non-negated occurrences of 'impossible'), and thus to every negative composite modal *tout court*. This finishes the proof that in Fabri's modal logic every composite modal constitutes an equivalence class with its co-existing quantifier mates.

To sum up: if, like Fabri, we maintain that *dicta* in composite modals function as common terms that supposit materially, and that the material *supposita* of such *dicta* are nothing but occurrences of the same proposition type, then we should accept that quantifying over *dicta* actually serves no purpose, as the quantifier is unable to leave its mark on the truth conditions of the proposition in question. We shall refer to this phenomenon as *quantification invariance*. Thus, the upshot of the above paragraphs is that composite modals in Fabri's modal logic are *invariant under quantification*. This is an important indication that Fabri's modal logic classifies as a *modernist* modal logic, as we shall see in Chap. 5.

There is no passage in the commentary on either *De int*. 12-13 or *An. pr*. I.3, 8-22 where Fabri explicitly states that composite modals are equivalent with their co-existing quantifier mates. But he is aware of this equivalence. This is proven by the passage in quote [*e*]. Fabri there explains how we should determine the truth value of composite modals. The procedure he introduces, relies on what he calls the 'reducing-to-assertoric' (*positio inesse*) principle. *Positio inesse* is the syntactic operation by which we identify the assertoric propositions that underlie complex structures such as hypothetical and modal propositions.[25] As Fabri explains at [i], in the case of composite modals *positio inesse* comes down to construing an underlying assertoric proposition of the modal proposition's *dictum*. The proposition so produced is a material *suppositum* of the *dictum*, and we determine the truth value of the composite modal by assessing whether this *suppositum* is among the *supposita* of the proposition's modality ([ii]).

> [*e*] Second, it is doubted how to judge of any modal proposition whether it is true or false. It is said that this is judged by the reducing-to-assertoric principle (*per positionem inesse*); that is, by reducing the modal proposition to an assertoric proposition of the present tense. The following rules apply to this principle. [i] The first rule is that composite modals are reduced to assertoric propositions by changing the accusatives of the *dictum* into nominatives, and

[25] In disputational contexts, '*ponere inesse*' means positing, or subscribing to, a proposition, as Fabri says in the questions on *An. pr*. I; see **S**, f. 241ra. The meaning of the term is also commented on by Soto; see Soto, *in Sum. log*. (Soto 1529: f. 65ra).

> the infinitive into a personal verb of the present tense and of the indicative mood. For instance, 'That something white is black is possible' (*album esse nigrum est possibile*) is reduced to an assertoric proposition as follows: 'Something white is black' (*album est nigrum*). [ii] And then we should see whether the modality occurring in the modal proposition can be truly predicated of (*possit uere predicari de*) the assertoric proposition so obtained. If so, then the modal proposition is true; if not, then false.[26]

Note that the *positio inesse* procedure can also, and perhaps equally plausibly, be interpreted in a much more radical way, namely as involving the *creation* of a new propositional token (which did not exist before the act of creation) rather than the *adduction* of an already existing propositional token, which is how we interpreted the procedure in the above paragraph. If we interpret *positio inesse* in this more radical fashion, then we commit Fabri to the claim that the very process of determining the truth value of a composite modal results in the proposition's satisfying existential import: on this alternative reading, every composite modal whose truth value is assessed at time t satisfies existential import at t; or, contrapositively: if a composite modal does not satisfy existential import at t, then its truth value has not been assessed at t. Thus, on this account, whether the truth value of a composite modal is determined can have an impact on the properties of the proposition, and also on its truth value. Take the proposition 'that every man is an animal is possible', and assume that it exists at t. If its *dictum* is empty at t, then the proposition is false and its truth value has not been determined. If the proposition's truth value at t has been determined, then its *dictum* is non-empty and the proposition is true (see the templates further on in this chapter). This conclusion is rather unsettling, since we prefer to think of truth and falsity as objective properties that are possessed by propositions irrespective of whether their truth value is analyzed or not. The radical interpretation cannot be ruled out, however. It is compatible with the textual evidence, just as the weaker interpretation that is proposed in the previous paragraph is.

Fabri does not differentiate between propositional quantities in quote [*e*], but instead describes one single procedure for determining the truth value of all composite modals, *regardless* of their quantity. We may summarize the procedure as follows:

> Given a composite modal φ with *dictum* d and modality m, produce a material *suppositum* ψ of d. If m does not have a negation sign attached to it, then φ is true iff ψ is among the *supposita* of m; while, if m has a negation sign attached to it, then φ is true iff ψ is not among the *supposita* of m.

where φ is a composite modal of *any* quantity, whether singular, indefinite, particular or universal. Is Fabri mistaken by not differentiating between propositional quantities? The answer is 'no'. By not differentiating, he makes sure that if the procedure is applied to a set of quantifier mates at t, then it will assign each member of this set the same truth value at t as all other members. For instance, suppose we were to assess

[26] **S**, ll. 427–434. See also **S**, f. 241[ra]: "[positio inesse] . . . tantum ualet sicut propositionem produci inesse."

the truth value of the universal proposition 'every that every man is running is possible' (*omne omnem hominem currere est possibile*) and its particular and singular quantifier mates at *t*. Then, by quote [*e*], we should produce three *supposita* – one for each proposition – by applying the 'reducing-to-assertoric' principle. The *supposita* produced are occurrences of 'every man is running', and they either all belong, or all do not belong, to the *supposita* of 'possible' at *t*. Thus the three propositions will be assigned the same truth value. That is, the procedure labels a set of co-existing quantifier mates as constituting one equivalence class; and this strongly suggests that it is developed starting from the observation that composite modals are invariant under quantification.

With quantification invariance introduced, we can finally reconstruct Fabri's truth conditions for composite modals. Let '$\times$' be a variable for either the universal, particular, indefinite or singular quantifier, and let '*m*' be a variable for any of the modalities in $\mathfrak{M}$. If *m* is 'necessary', 'possible', or 'contingent', then, by Fabri's light, propositions of the form '$\times$ *d* is *m*' are true just in case there is a *suppositum* of *d* that is among the *supposita* of *m*; whereas if *m* is 'impossible', then propositions of the form '$\times$ *d* is *m*' are true just in case *either d* is empty *or* there is a *suppositum* of *d* that is among the *supposita* of *m*. We can further refine these truth conditions by taking into account the supposition of the different modalities. If we do this, then we obtain the equivalences in the table below, which give the truth conditions for affirmative composite modals of necessity, possibility and contingency, and for negative composite modals of impossibility. They are formulated in the symbolism introduced in Chap. 1. As mentioned there, the expression '(ψ)' denotes the *dictum* corresponding to the assertoric proposition ψ. If 'ψ' occurs underlined (i.e., '$(\underline{\psi})$'), then the modal proposition in which '$(\underline{\psi})$' occurs is true only if the *dictum* has at least one *suppositum* (i.e., is non-empty). The one-sided reading of 'contingent' is marked '∇_1', and the two-sided reading '∇_2'. Note that only the truth conditions for indefinite composite modals are incorporated in the table. Due to quantification invariance these conditions equally apply to the quantifier mates of these propositions.

- $\Box(\underline{\psi})$, if expressed at time *t*, is true at *t* iff there is at least one *suppositum* ψ at *t*, and for every *t'*, if a proposition ψ exists at *t'*, then it is true at *t'*
- $\Diamond(\underline{\psi})$, if expressed at time *t*, is true at *t* iff there is at least one *suppositum* ψ at *t* and for some *t'*, if a proposition ψ exists at *t'*, then it is true at *t'*
- $\bigcirc(\psi)$, if expressed at time *t*, is true at *t* iff there is no *suppositum* ψ at *t* or for every *t'*, if a proposition ψ exists at *t'*, then it is not true at *t'*
- $\nabla_1(\underline{\psi})$, if expressed at time *t*, is true at *t* iff $\Diamond(\underline{\psi})$ is true at *t*
- $\nabla_2(\underline{\psi})$, if expressed at time *t*, is true at *t* iff there is at least one *suppositum* ψ at *t*, and for some *t'*, *t"*, if a proposition ψ exists at *t'*, then it is true at *t'*, and if a proposition ψ exists at *t"*, then it is not true at *t"*

In conclusion to this chapter, it is useful to again summarize the main elements from our discussion. This is done in statements $\mathbf{A}_{4–7}$, which should be read in connection with statements $\mathbf{A}_{1–3}$ from the end of Chap. 2.

- $\mathbf{A}_4$: A proposition is a *composite modal* if it is an instance of one of the templates in T_{CM}.
- $\mathbf{A}_5$: A composite modal is equivalent to its quantifier mates with which it co-exists.
- $\mathbf{A}_6$: An *affirmative* composite modal of necessity, possibility or contingency is true iff there is a *suppositum* of its *dictum* that is among the *supposita* of its modality.
- $\mathbf{A}_7$: A *negative* composite modal of necessity, possibility or contingency is true iff either its *dictum* is empty or there is a *suppositum* of its *dictum* that is not among the *supposita* of its modality.

References

Ashworth EJ (1974) Language and logic in the post-medieval period. Reidel, Dordrecht

Ashworth EJ (1978) Theories of the proposition. Some early sixteenth-century discussions. Franciscan Studies 38:81–121

Ashworth EJ (1985) Studies in post-medieval semantics. Variorum Reprints, London

Ashworth EJ (1990) The doctrine of signs in some early sixteenth-century Spanish logicians. In: Angelelli I, d'Ors A (eds) Estudios de historia de la lógica. Eunate, Pamplona, pp 13–38

Ashworth EJ (2013) Descent and ascent from Ockham to Domingo de Soto. An answer to Paul Spade. Vivarium 51:385–410

Becker O (1952) Untersuchungen über den Modalkalkül. Anton Hain, Meisenheim am Glan

Binini I (2022) Possibility and necessity in the time of Peter Abelard. Brill, Leiden

Bursil-Hall GL (1971) Speculative grammars of the Middle Ages. The doctrine of 'Partes Orationis' of the Modistae. Mouton, The Hague

Chellas BF (1980) Modal logic. An introduction. Cambridge University Press, Cambridge

Coronel A (1517) Prima pars rosarii. Venalia apud Oliuerium Senant, [Paris]

Ebbesen S (2019) Imposition of words in Stoicism and late ancient grammar and philosophy. *Methodos, 19* (online journal). https://doi.org/10.4000/methodos.5641

Hughes GE, Cresswell MJ (1996) A new introduction to modal logic. Routledge, London

Karger E (1982) La supposition materielle comme supposition significative: Paul de Venice, Paul de Pergula. In: Maierù A (ed) English logic in Italy in the 14th and 15th centuries. Bibliopolis, Napoli, pp 331–341

Klima G (2009) John Buridan. Oxford University Press, Oxford

Knuuttila S (1993) Modalities in medieval philosophy. Routledge, New York

Kretzmann N (1970) Medieval logicians on the meaning of the *propositio*. J Philos 67:767–787

Pérez-Ilzarbe P (2004) John Buridan and Jerónimo Pardo on the notion of *propositio*. In: Friedman RL, Ebbesen S (eds) John Buridan and beyond. Topics in the language sciences, 1300–1700. C.A. Reitzel, København, pp 153–181

Read S (1999) How is material supposition possible? Mediev Philos Theol 8:1–20

Soto D (1529) Summule. In officina Joannis Junte, Burgis

Spade PV (1975) Ockham's distinctions between absolute and connotative terms. Vivarium 13:55–76

Chapter 4
Model-Theoretic Reconstruction of Fabri's Logic of Composite Modals

Abstract This chapter gives a rational reconstruction of Fabri's logic of composite modals. The main aim of the chapter is to get clear on the token-based nature of this logic. We develop two model-theoretic semantics which both do justice to Fabri's original account. Based on these semantics we argue that Fabri adopted a moderate version of the token-based approach, and avoided most issues related to propositional existence that usually come with this approach.

Keywords Rational reconstruction · Modal logic · Composite modals · Token-based semantics

In Chap. 3 we took an important step to get a grasp of Fabri's logic of composite modals. But as the reader might have noticed, the discussion remained slightly elusive. Fabri adopts a reductionist approach where modalities are interpreted extensionally in a temporal framework.[1] Yet Fabri never really discusses the properties of this framework, and in Chap. 3 we made no attempt to go beyond Fabri's text. We will take up the challenge in this chapter, and try to get a grasp of the nature of the temporal structures on which Fabri's logic of composite modals is interpreted. A good way to make progress on the issue is to consider a small set of implications involving composite modals that Fabri considers valid, and to construct a model-theoretic framework that validates each of these implications. The properties of this framework, so the hypothesis goes, might tell us something more about the properties of Fabri's original. We will be concerned with the following seven implications,

[1] Fabri's logic is similar in this regard to the various modal semantics that were developed from the 1950s onwards, starting with Kripke's possible world semantics. For a historical overview of the evolution of modal logic during the twentieth century, see esp. Copeland (2002); Gabbay and Woods (2006).

C. Geudens, L. Demey, *The Modal Logic of John Fabri of Valenciennes (c. 1500)*, SpringerBriefs in Philosophy, https://doi.org/10.1007/978-3-030-98802-9_4

which it seems any reasonable logic of composite modals should include, and which are indeed held by Fabri.[2]

(a) If $\Box(\underline{\psi})$ is true, then ψ is true

(b) If $\Box(\underline{\psi})$ is true, then $\Diamond(\underline{\psi})$ is true

(c) If $\Box(\underline{\psi})$ is true, then $\neg\Diamond(\neg\psi)$ is true

(c*) If $\neg\Diamond(\neg\psi)$ is true, then $\Box(\underline{\psi})$ is true

(d) If $\Diamond(\underline{\psi})$ is true, then $\neg\Box(\neg\psi)$ is true

(d*) If $\neg\Box(\neg\psi)$ is true, then $\Diamond(\underline{\psi})$ is true

(e) If $\Box(\underline{\psi_1 \rightarrow \psi_2})$ and $\Box(\underline{\psi_1})$ are both true, then $\Box(\underline{\psi_2})$ is also true

In (c), (d) and their converses (c*) and (d*), '$\neg\psi$' is the contradictory negation of ψ. Thus, if '(ψ)' is 'That every man is running', then '$(\neg\psi)$' is 'That not every man is running'. The reader will have noticed that each of these implications resembles a well-known axiom from present-day system T.[3]

Principles (a)–(e) constitute a solid basis for a formal reconstruction of Fabri's logic of composite modals. However, by themselves they do not allow us to uniquely pin down one single reconstruction. In particular, in this chapter we develop two model-theoretic semantics which both validate principles (a)–(e) and thus both seem faithful representations of Fabri, but which are nevertheless quite different from each other. We presently do not have a definitive argument for either semantics as the unique best representation of Fabri's thought. However, we will see that both frameworks do have enough features in common to allow us to reach some robust conclusions about the nature of Fabri's token-based approach to composite modals.

Both formalizations will make use of two unary metalinguistic predicates, one for existence (E) and one for truth (T). The existence predicate is needed to account for the token-based nature of Fabri's logic – or so it seems at first sight. Further on we

---

[2]For (a), see **S**, f. 238va: "... primo modo sillogizando ex modalibus compositis sillogizatur eodem modo sicut ex illis de inesse ... exemplum in prima figura: 'Omne necessarium est uerum; omnem hominem esse animal est necessarium; ergo omne omnem hominem esse animal est uerum', et regulatur iste sillogismus per dici de omni". For (b), see **S**, ll. 332–426. For (c), see **S**, ll. 286–297 (see also quote [*d*]). For its converse, (c*), see **S**, ll. 303–307. For (d), see **S**, ll. 300–303. For (e), see **S**, ff. 231ra: "... fundatur in ista maxima: 'si antecedens bone consequentie est necessarium, consequens est necessarium'"; as well as **S**, f. 238vb: "... bene ualet: maior de Barbara est necessaria et minor de Barbara est necessaria, ergo et conclusio de Barbara est [ergo conclusio de Barbara est; *rep.* **S**] necessaria ... Preterea talis modus arguendi fundatur in ista maxima: 'si antecedens est necessarium, consequens est necessarium'." Finally, the converse of (d), i.e. (d*), is not expressly stated in Fabri's text, but Fabri does say that two composite modals involving a modality from the same vertex in the modal square of opposition are equivalent to each other, which implies that he accepts the equivalence of affirmative composite modals of possibility with negative composite modals of necessity, and thus also the validity of (d*); see **S**, ll. 290–292. For more on the modal square, see Sect. 6.2.

[3]The resemblances are as follows: (a) and the T (or M) axiom, (b) and the D axiom, (c-d*) and the duality between necessity and possibility; (e) and the K (or distribution) axiom. On the modern system T, see, e.g., Hughes and Cresswell (1996: 41–43).

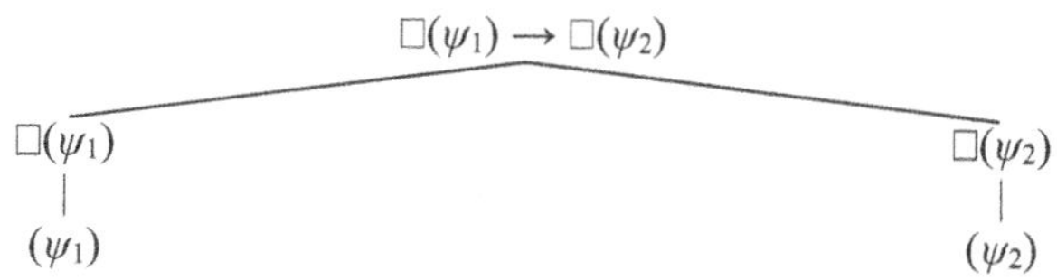

Fig. 4.1 The construction tree of $\Box(\psi_1) \rightarrow \Box(\psi_2)$. (A case could be made that since (ψ_1) and (ψ_2) are the nominalizations of resp. ψ_1 and ψ_2, these two propositions are subpropositions of (ψ_1) and (ψ_2), and thus should also occur in the tree. This would mean that the tree in its current form is incomplete, and, in light of principle (E), it would also mean that the existence of a *dictum* at some time t implies the existence, at t, of an underlying assertoric proposition of that *dictum*, which would in turn imply that *dicta* in composite modals satisfy existential import by definition. The worry is legitimate, but we should not decide on the issue here, as it is irrelevant for the proofs below; that is, none of the proofs hinges on the issue whether $E(\psi)$ implies $E\psi$)

shall see that on at least one plausible interpretation of Fabri, the existence predicate can actually be dispensed with. In both semantics, the E predicate has the following significant property:

(EC) For each proposition φ and time t, if $E\varphi$ is true at t, then $E\varphi'$ is true at t, for any subproposition φ' of φ.

For instance, if $E(\Box(\underline{\psi}_1) \rightarrow \Box(\underline{\psi}_2))$ is true at t, then so are $E(\Box(\underline{\psi}_1))$, $E(\Box(\underline{\psi}_2))$, $E(\psi_1)$ and $E(\psi_2)$, given the construction tree of $\Box(\underline{\psi}_1) \rightarrow \Box(\underline{\psi}_2)$ in Fig. 4.1.[4]

There is no passage in Fabri's text where (EC) is stated explicitly, but it is safe to assume that he would have accepted it. After all, propositions in token-based systems like Fabri's are items in the *res extra*. They are aggregates that are decomposable into terms, which, in turn, are decomposable into letters/sounds. All these constituents, moreover, are discrete items which together constitute a quantity; and while the expression 'This is a proposition' is true when pointing at the aggregate, it is false when pointing at its constituents. Decomposability and the discrete nature of the constituents, along with the fact that the name of the aggregate cannot be verified of the constituents, are all typical characteristics of the mereological structure that in medieval texts is usually called an 'integral whole' (*totum integrale*). That is, it can be plausibly maintained that, from a mereological point of view, propositions in token-based systems are integral wholes; and, as studies by Arlig and Henry have amply shown, one of the most widely accepted principles in medieval mereology holds precisely that an integral whole exists only if each of its constituent parts exists.[5] Thus, (EC) can be considered as an instance of this mainstream mereological

[4]Notice that whenever the arguments of E and T are complex symbol strings (i.e., strings consisting of at least one variable and a connective), like '$\neg\varphi$' or '$\Box(\underline{\varphi})$'), brackets are placed around them to indicate the scope of E and T. Thus, these brackets do not indicate nominalizations. For instance, while $E(\varphi)$ reads 'that φ exists', $E(\neg\varphi)$ reads '$\neg\varphi$ exists' rather than 'that $\neg\varphi$ exists'.

[5]See Arlig (2019) and the literature cited there; Henry (1991).

principle, meaning it is fairly uncontroversial to assume that Fabri would indeed have endorsed it.

As for the truth predicate, notice that Fabri accepts the so-called Rule of Contradictory Pairs, which states that two contradictory propositions satisfy the laws of Non-Contradiction and Excluded Middle, i.e. at least one of them should be true, yet they cannot be true together.[6] Thus, in particular, when a proposition φ is true at a time t, then its contradictory $\neg\varphi$ is not true at t. Bearing in mind that propositions are tokens and can thus only be true or false if they exist, we obtain that the truth predicate should satisfy the following condition, which once again holds in both semantics that we will develop.

(TC) For each proposition φ and time t, $T\varphi$ is true at t only if $E\varphi$ and $E(\neg\varphi) \rightarrow \neg T(\neg\varphi)$ are both true at t.

A further question is: how do we describe the case where φ would be true at t if it exists at t, but it is unknown whether φ exists at t in the first place? The case is a rather important one, as we shall see, so we need a term for it. Since Fabri does not seem to have one, we shall coin one ourselves, and say that φ *holds* at t in that case.[7]

(H) φ holds at t iff $E\varphi \rightarrow T\varphi$ is true at t.

It bears emphasizing that (H) is a metalinguistic principle. Formally, that $E\varphi \rightarrow T\varphi$ is true at t would be represented as $t \vDash E\varphi \rightarrow T\varphi$ (or t models $E\varphi \rightarrow T\varphi$) rather than as $t \vDash T(E\varphi \rightarrow T\varphi)$. The matter is important, since on the latter reading the holding-at relation would collapse into the truth relation. To see this, notice that $t \vDash T(E\varphi \rightarrow T\varphi)$ only if $t \vDash E(E\varphi \rightarrow T\varphi)$, by (TC). Thus, $t \vDash E\varphi$, by (EC), and so $t \vDash T\varphi$, by *modus*

[6] See, e.g., **S**, ll. 300–303, 303–307. Both passages contain a *reductio* proof. In the first proof, Fabri derives a contradiction from the assumption that two contradictory propositions are true together; in the second one, from the assumption that two contradictory propositions are false together. See Sect. 6.2 for further discussion of Fabri's account of the opposition relations. The label 'Rule of Contradictory Pairs' is taken from the scholarship on Aristotle's logic, where it was introduced by Whitaker (1996). On the medieval reception of this Aristotelian principle, see Read (2020a, b) and the literature cited there.

[7] The differentiation between truth and holding-at is key to the token-based approach, and it is thus unsurprising that several alternative proposals are to be found in the scholarship. To give but one example, as part of a model-theoretic reconstruction of the (token-based) account of consequence (*consequentia*) of John Buridan, Dutilh Novaes (2005) differentiates between 'context of formation' and 'context of evaluation'. The former context addresses the existence criterion, while the latter context addresses the holding-at criterion; and a proposition, which is interpreted as a function from formation-evaluation pairs to the truth value set, maps to the value 'true' just in case both contexts are satisfied.

ponens. That is, if "φ holds at t" is read as $t \vDash T(E\varphi \rightarrow T\varphi)$, then φ holds at t only if φ is true at t, and the concepts of truth and holding-at would be conflated.[8]

With (EC), (TC) and (H) introduced, we can proceed to our two formal reconstructions, which will be called *ML* and *ML'*. Both systems work with the same language $\mathcal{L}$, whose well-formed formulas are generated by applying connectives and predicates to a set *AtProp* of atomic propositions, which are here to be understood as affirmative assertoric propositions of any quantity, and either singular, particular, indefinite or universal. The syntax of $\mathcal{L}$ is defined as follows:

$$\varphi ::= p \mid \neg\varphi \mid \varphi \wedge \psi \mid \varphi \vee \psi \mid \varphi \rightarrow \psi \mid E\varphi \mid T\varphi \mid \Box(\underline{\varphi}) \mid \Diamond(\underline{\varphi})$$

where $p \in AtProp$. In the first formal system, *ML*, the language $\mathcal{L}$ is interpreted on models $\mathbb{M} = \langle W, R, V, ex \rangle$. The components of these models are defined as follows.

- W is a set of possible times t, where a time t is possible only if for each pair of contradictory propositions $\{\varphi, \neg\varphi\}$, it is not the case that both φ and $\neg\varphi$ hold at t. Using the semantics of $\mathcal{L}$ that is introduced immediately below, this condition can be formally expressed as follows:

(PT) $\mathbb{M}, t \nvDash (E\varphi \rightarrow T\varphi) \wedge (E(\neg\varphi) \rightarrow T(\neg\varphi))$, for every $\mathbb{M}$, t.

- R is a binary accessibility relation on W, i.e., $R \subseteq W \times W$. Note that Fabri accepts the validity of (a); that is, he holds that the truth of a composite modal of necessity at time t implies the truth of the *dictum*'s underlying assertoric propositions at t. That means that we should impose a reflexivity condition on R, meaning tRt, for every $t \in W$.
- V: $W \rightarrow \wp(AtProp)$ is a valuation function, specifying which atomic propositions are true at which times. The value $V(t)$ of a time t is the set of atomic propositions that are true at t.
- ex: $W \rightarrow \wp(\mathcal{L})$ is an existence function, specifying which propositions exist at which times. The value $ex(t)$ of a time t is the set of propositions, whether atomic or not, existing at t. We assume that $ex(t)$ is closed under subpropositions, i.e., if $\varphi \in ex(t)$, then $\varphi' \in ex(t)$, for all subpropositions φ' of φ.

The semantics of $\mathcal{L}$ is as follows:

	$\mathbb{M}, t \vDash p$	iff	$p \in V(t)$
$(\neg)$	$\mathbb{M}, t \vDash \neg\varphi$	iff	$\mathbb{M}, t \nvDash \varphi$

[8] We thank an anonymous reviewer for pointing us to this issue.

$(\wedge)$	$\mathbb{M}, t \vDash \varphi \wedge \psi$	iff	$\mathbb{M}, t \vDash \varphi$ and $\mathbb{M}, t \vDash \psi$
$(\vee)$	$\mathbb{M}, t \vDash \varphi \vee \psi$	iff	$\mathbb{M}, t \vDash \varphi$ or $\mathbb{M}, t \vDash \psi$
$(\rightarrow)$	$\mathbb{M}, t \vDash \varphi \rightarrow \psi$	iff	$\mathbb{M}, t \nvDash \varphi$ or $\mathbb{M}, t \vDash \psi$
(E)	$\mathbb{M}, t \vDash E\varphi$	iff	$\varphi \in ex(t)$
(T)	$\mathbb{M}, t \vDash T\varphi$	iff	$\mathbb{M}, t \vDash E\varphi \wedge \varphi$
$(\Box)$	$\mathbb{M}, t \vDash \Box(\underline{\varphi})$	iff	$\mathbb{M}, t \vDash E\varphi$ and for all t' s.t. tRt', $\mathbb{M}, t' \vDash E\varphi \rightarrow T\varphi$
$(\Diamond)$	$\mathbb{M}, t \vDash \Diamond(\underline{\varphi})$	iff	$\mathbb{M}, t \vDash E\varphi$ and for some t' s.t. tRt', $\mathbb{M}, t' \vDash E\varphi \rightarrow T\varphi$

It is worth noticing that (EC) holds in *ML*, following the assumption that $ex(t)$ is closed under subpropositions. This principle, along with (T), also establishes the following two theorems about *ML*, as the reader is invited to check.

(A) For every $\mathbb{M}$, t, if $\mathbb{M}, t \vDash E(\neg\varphi)$, then $\mathbb{M}, t \vDash \neg T\varphi$ just in case $\mathbb{M}, t \vDash T(\neg\varphi)$
(B) For every $\mathbb{M}$, t, if $\mathbb{M}, t \vDash T(\neg\varphi)$, then $\mathbb{M}, t \vDash \neg T\varphi$

That is, under the assumption of $E(\neg\varphi)$ (and thus also of $E\varphi$, by (EC)), the falsity of a proposition is equivalent to the truth of the negation of that proposition (A). Furthermore, the truth of a negation $\neg\varphi$ implies the falsity of the negation's corresponding affirmation φ (B). The principle (TC), too, is universally valid in *ML*. Here is a proof.

$$\text{(TC)}\quad \mathbb{M}, t \vDash T\varphi \rightarrow (E\varphi \wedge (E(\neg\varphi) \rightarrow \neg T(\neg\varphi))), \text{ for every } \mathbb{M}, t.$$

Proof To show that $\mathbb{M}, t \vDash T\varphi \rightarrow (E\varphi \wedge (E(\neg\varphi) \rightarrow \neg T(\neg\varphi)))$, for every $\mathbb{M}$, t, it suffices to show that (1) $\mathbb{M}, t \vDash T\varphi \rightarrow E\varphi$ and (2) $\mathbb{M}, t \vDash T\varphi \rightarrow (E(\neg\varphi) \rightarrow \neg T(\neg\varphi))$, for every $\mathbb{M}$, t. Consider an arbitrary time t in a model $\mathbb{M}$ and assume that $\mathbb{M}, t \vDash T\varphi$. To prove (1), notice that $\mathbb{M}, t \vDash E\varphi \wedge \varphi$, by ($T$), and thus also $\mathbb{M}, t \vDash E\varphi$, by ($\wedge$). To prove (2), further assume that $\mathbb{M}, t \vDash E(\neg\varphi)$. By (A) from above it follows that $\mathbb{M}, t \vDash \neg T\varphi$ iff $\mathbb{M}, t \vDash T(\neg\varphi)$, or equivalently, $\mathbb{M}, t \vDash T\varphi$ iff $\mathbb{M}, t \vDash \neg T(\neg\varphi)$. Since indeed $\mathbb{M}, t \vDash T\varphi$, it follows that also $\mathbb{M}, t \vDash \neg T(\neg\varphi)$. ∎

Furthermore, Fabri's truth conditions of composite modals of possibility and necessity, which we reconstructed earlier, can be faithfully represented in *ML*:

$\mathbb{M}, t \vDash T(\Box(\underline{\varphi}))$	iff	$\mathbb{M}, t \vDash E(\Box(\underline{\varphi})) \wedge \Box(\underline{\varphi})$	
		$\mathbb{M}, t \vDash E(\Box(\underline{\varphi}))$ and $\mathbb{M}, t \vDash \Box(\underline{\varphi})$	by $(\wedge)$
		$\mathbb{M}, t \vDash E(\Box(\underline{\varphi}))$ and $\mathbb{M}, t \vDash E\varphi$ and for all t' such that tRt', $\mathbb{M}, t' \vDash E\varphi \rightarrow T\varphi$	by $(\Box)$
$\mathbb{M}, t \vDash T(\Diamond(\underline{\varphi}))$	iff	$\mathbb{M}, t \vDash E(\Diamond(\underline{\varphi})) \wedge \Diamond(\underline{\varphi})$	
		$\mathbb{M}, t \vDash E(\Diamond(\underline{\varphi}))$ and $\mathbb{M}, t \vDash \Diamond(\underline{\varphi})$	by $(\wedge)$
		$\mathbb{M}, t \vDash E(\Diamond(\underline{\varphi}))$ and $\mathbb{M}, t \vDash E\varphi$ and for some t' such that tRt', $\mathbb{M}, t' \vDash E\varphi \rightarrow T\varphi$	by $(\Diamond)$

This, along with the fact that (TC) and (EC) are both valid in *ML*, suggests that the structure of the models $\mathbb{M}$ on which $\mathcal{L}$ is interpreted, approximates the temporal structure that is presupposed by Fabri. This is further corroborated by the fact that the implications (a')-(e'), which are the model-theoretic translations of the formulas (a)-(e), are universally valid in *ML*, provided the existence of the argument of the T predicate in the consequent is assumed at all relevant times. The proofs are given in Appendix II.

(a') $T(\Box(\underline{\psi})) \rightarrow T\psi$
(b') $T(\Box(\underline{\psi})) \rightarrow T(\Diamond(\underline{\psi}))$
(c') $T(\Box(\underline{\psi})) \rightarrow T(\neg\Diamond(\neg\psi))$
(c*') $T(\neg\Diamond(\neg\psi)) \rightarrow T(\Box(\underline{\psi}))$
(d') $T(\Diamond(\underline{\psi})) \rightarrow T(\neg\Box(\neg\psi))$
(d*') $T(\neg\Box(\neg\psi)) \rightarrow T(\Diamond(\underline{\psi}))$
(e') $T(\Box(\underline{\psi_1 \rightarrow \psi_2})) \rightarrow (T(\Box(\underline{\psi_1})) \rightarrow T(\Box(\underline{\psi_2})))$

To summarize: *ML* seems to be a faithful representation of Fabri's account of composite modals. We can thus get a better grasp of the nature of this account by focusing on the properties of *ML*. Remember that following Fabri's lead we stipulated that any time where the conjunction of $E\varphi \rightarrow T\varphi$ and $E(\neg\varphi) \rightarrow T(\neg\varphi)$ is true (i.e., at which both φ and $\neg\varphi$ hold), is impossible. The implication in both these conjuncts is the material implication, and that means that we draw in the paradoxes of material implication, viz. *ex falso quodlibet* and *verum ex quolibet*. It follows, in particular, that the principle (IT) holds for *ML*-models (where 'IT' is short for 'Impossible Time'):

(IT) There is no *ML*-model $\mathbb{M}$ and time t such that $\mathbb{M}, t \vDash \neg E\varphi \wedge \neg E(\neg\varphi)$.

Proof Assume, towards a contradiction, that there is a time t in an *ML*-model $\mathbb{M}$ such that $\mathbb{M}, t \vDash \neg E\varphi \wedge \neg E(\neg\varphi)$. Then also $\mathbb{M}, t \vDash \neg E\varphi$ and $\mathbb{M}, t \vDash \neg E(\neg\varphi)$, by ($\wedge$). From $\mathbb{M}, t \vDash \neg E\varphi$ we obtain $\mathbb{M}, t \vDash E\varphi \rightarrow T\varphi$, by ($\neg$) and ($\rightarrow$). By parity of reasoning, we obtain $\mathbb{M}, t \vDash E(\neg\varphi) \rightarrow T(\neg\varphi)$ from $\mathbb{M}, t \vDash \neg E(\neg\varphi)$. Thus also $\mathbb{M}, t \vDash (E\varphi \rightarrow T\varphi) \wedge (E(\neg\varphi) \rightarrow T(\neg\varphi))$, by ($\wedge$). But this is impossible, since $\mathbb{M}, t \nvDash (E\varphi \rightarrow T\varphi) \wedge (E(\neg\varphi) \rightarrow T(\neg\varphi))$, by (PT). ∎

Informally, at any possible time either a proposition exists or its negation exists. Possible times, it turns out, are precisely those times where φ exists, for each proposition φ. Call this the (ERed) theorem about *ML*.

(ERed) For every *ML*-model $\mathbb{M}$, time t and proposition φ: $\mathbb{M}, t \vDash E\varphi$.

Proof Let t be an arbitrary time in an *ML*-model $\mathbb{M}$, and let φ be an arbitrary proposition in $\mathcal{L}$. It follows from (IT) that $\mathbb{M}, t \vDash E\varphi \vee E(\neg\varphi)$, where. But $\mathbb{M}, t \vDash E(\neg\varphi)$ only if $\mathbb{M}, t \vDash E\varphi$, by (EC). Thus, $\mathbb{M}, t \vDash E\varphi \vee E(\neg\varphi)$ just in case $\mathbb{M}, t \vDash E\varphi$. ∎

That is, all truth conditions in *ML* are formulated in such a way that truth is made conditional upon propositional existence. But if we study *ML* in detail, then it turns out that the existence conditions are actually satisfied by definition. Adding these conditions is thus redundant (hence the name 'ERed'), and the existence predicate E can in principle be dispensed with.[9] For instance, again consider the semantics for $T\varphi$, $\neg T\varphi$, $T(\neg\varphi)$, and $T(\Box(\underline{\varphi}))$.

$$\begin{array}{lll}
\mathbb{M}, t \vDash T\varphi & \text{iff} & \mathbb{M}, t \vDash E\varphi \wedge \varphi \\
\mathbb{M}, t \vDash \neg T\varphi & \text{iff} & \mathbb{M}, t \vDash \neg E\varphi \vee \neg\varphi \\
\mathbb{M}, t \vDash T(\neg\varphi) & \text{iff} & \mathbb{M}, t \vDash E(\neg\varphi) \wedge \neg\varphi \\
\mathbb{M}, t \vDash T(\Box(\underline{\varphi})) & \text{iff} & \mathbb{M}, t \vDash E(\Box(\underline{\varphi})) \text{ and } \mathbb{M}, t \vDash E\varphi \text{ and} \\
& & \text{for all } t' \text{ such that } tRt', \mathbb{M}, t' \vDash E\varphi \rightarrow T\varphi
\end{array}$$

Since $\mathbb{M}, t \vDash E\varphi$ by (ERed), we have that $\mathbb{M}, t \vDash E\varphi \wedge \varphi$ is equivalent with $\mathbb{M}, t \vDash \varphi$, and so $\mathbb{M}, t \vDash T\varphi$ is equivalent with $\mathbb{M}, t \vDash \varphi$, while $\mathbb{M}, t \vDash \neg T\varphi$ is equivalent with $\mathbb{M}, t \vDash \neg\varphi$. Likewise $\mathbb{M}, t \vDash E(\neg\varphi)$ by (ERed), and so $\mathbb{M}, t \vDash T(\neg\varphi)$ is equivalent with $\mathbb{M}, t \vDash \neg\varphi$, just like $\mathbb{M}, t \vDash \neg T\varphi$. We thus arrive at the rather intuitive idea (from a modern, Boolean perspective, at least) that the falsity (or: non-truth) of a proposition is equivalent to the truth of the negation of that proposition. By the same reasoning, we also obtain that $\mathbb{M}, t \vDash T(\Box(\underline{\varphi}))$ just in case for all t' such that tRt', $\mathbb{M}, t' \vDash \varphi$, which comes close to stating the modern (non-token-based) idea of necessity as truth at all possible worlds. Furthermore, with (ERed) available, we can see that (EC) can actually be derived in *ML* and thus does not need to be assumed independently. We do not need to assume the existence of the argument of the T predicate in the consequent to establish the universal validity of (a')-(e') in *ML* either – the existence condition is satisfied by definition.

The formal framework of *ML*, and the (ERed) theorem in particular, teach us an important lesson about Fabri's account of composite modals. Even though throughout his discussion Fabri occasionally emphasizes that propositional existence is a necessary condition for both truth and validity (see, e.g., quote [*d*]), his logic turns out to be so conceived that the propositional existence requirement is satisfied by definition: at least one token of each proposition type exists at every time. Note that this has significant consequences for the existential import of *dicta* in composite modals, among other things. Remember that Fabri says that the non-existence of a

[9]On a semantic level, (ERed) means that for every *ML*-model $\mathbb{M} = \langle W, R, V, ex\rangle$ and time t, it holds that $ex(t) = \mathcal{L}$. (Note that $ex(t)$ is thus *a fortiori* closed under subpropositions, as required.) In other words, the *ex*-function is constant, and is in a sense thus redundant, just like the existence predicate E that it provides the semantics of.

material *suppositum* of the *dictum* is sufficient for the falsity of an affirmative composite modal, and for the truth of a negative composite modal. It turns out that these conditions are also redundant, as (the *ML* formalization of) Fabri's system does not allow the case where the material supposition of the *dictum* is empty. There is no possible time where the falsity of an affirmative composite modal is due to the emptiness of the *dictum*, just as there is no possible time where the truth of a negative composite modal is due to the emptiness of the *dictum*.

This concludes our discussion of *ML* as a formal reconstruction of Fabri. Our second system, *ML'*, will be introduced in two steps: we first introduce the system ML^0, which will subsequently be expanded to the final system *ML'*. Let us start with the system ML^0. This has precisely the same language $\mathcal{L}$ and precisely the same semantic clauses as *ML*. Furthermore, it also works with models $\mathbb{M} = \langle W, R, V, ex \rangle$, where the components R, V and ex are exactly as before. The only difference between *ML* and ML^0 lies in the characterization of 'possible times', i.e., the definition of the set W. Recall that in *ML*, a time t is said to be possible only if for each pair of contradictory propositions $\{\varphi, \neg\varphi\}$, it is not the case that both φ and $\neg\varphi$ hold at t. Formally, this was expressed by (PT).

(PT) $\mathbb{M}, t \nvDash (E\varphi \rightarrow T\varphi) \wedge (E(\neg\varphi) \rightarrow T(\neg\varphi))$, for every *ML*-model $\mathbb{M}$ and time t.

However, (PT) is one of the few aspects of *ML* that cannot be explicitly traced back to Fabri's text. It is merely a plausible interpretation of the concept of possible time in token-based settings in general; and it is by no means the *only* plausible interpretation of this concept. In particular, the new system ML^0 is based on the assumption that possible times should be interpreted in terms of the notion of *truth*, rather than the notion of *holding-at*. On this approach, a time t is possible only if a proposition and its negation are not both true at t in case they co-exist at t. Call this alternative (PT').

(PT') $\mathbb{M}, t \vDash (E\varphi \wedge E(\neg\varphi)) \rightarrow \neg(T\varphi \wedge T(\neg\varphi))$, for every ML^0-model $\mathbb{M}$ and time t.

It is easy to check that ML^0 is a strictly weaker semantics than *ML*, since (PT) entails (PT'), but not *vice versa*.[10] Or put differently: every *ML*-model is also an ML^0-model, but not *vice versa*. There thus exist ML^0-models that are not *ML*-models. In particular, there exist ML^0-models $\mathbb{M} = \langle W, R, V, ex \rangle$ and times t such that $ex(t)$ is *not* the entire language $\mathcal{L}$ (a concrete example will be provided below), which violates (ERed) and thus immediately means that $\mathbb{M}$ is not an *ML*-model.

[10]To go from (PT) to (PT'), assume that $\mathbb{M}, t \nvDash (E\varphi \rightarrow T\varphi) \wedge (E(\neg\varphi) \rightarrow T(\neg\varphi))$. By ($\wedge$) it follows that $\mathbb{M}, t \nvDash E\varphi \rightarrow T\varphi$ or $\mathbb{M}, t \nvDash E(\neg\varphi) \rightarrow T(\neg\varphi)$. Let's assume that $\mathbb{M}, t \nvDash E\varphi \rightarrow T\varphi$; the other case is completely similar. By ($\rightarrow$) it follows that $\mathbb{M}, t \nvDash T\varphi$, and by ($\neg$) and ($\wedge$) it follows that $\mathbb{M}, t \vDash \neg(T\varphi \wedge T(\neg\varphi))$. Finally, by ($\rightarrow$) it follows that $\mathbb{M}, t \vDash (E\varphi \wedge E(\neg\varphi)) \rightarrow \neg(T\varphi \wedge T(\neg\varphi))$, as desired.

Just like the original semantics *ML*, the new semantics ML^0 validates the principles (EC) and (TC). After all, the proofs of these principles in *ML* did not rely on (PT), and are thus unaffected by the change to (PT') in ML^0. Completely similarly, ML^0 validates principle (a') as well: the proof of this principle (in Appendix II) does not rely on (PT), and thus continues to hold in ML^0. By contrast, ML^0 does *not* validate principles (b')-(e'): the proofs of these principles in Appendix II crucially rely on (ERed), and thus indirectly on (PT). Let's sketch a concrete counterexample against (b'); counterexamples against (c')-(e') are equally straightforward to formulate. Define an ML^0-model $\mathbb{M} = \langle W, R, V, ex \rangle$ with $W = \{t, t'\}$, $R = W^2$, $ex(t) = \{\Box(\underline{p}), p\}$ and $ex(t') = \{p\} = V(t) = V(t')$. It is now easy to check that $\mathbb{M}$, $t \vDash T(\Box(\underline{p}))$ and yet $\mathbb{M}, t \nvDash T(\Diamond(\underline{p}))$, which means that principle (b') is invalidated at $\mathbb{M}$, t. Note that this counterexample crucially relies on the fact that $ex(t)$ does not contain the proposition $\Diamond(\underline{p})$, or more generally, on the fact that $ex(t)$ is not the entire language (which means that our $\mathbb{M}$ is an ML^0-model but *not* an *ML*-model).

Let us take stock. The system ML^0 is closely related to *ML*, but starts from a different characterization of 'possible times', viz. in terms of the notion of *truth*, rather than the notion of *holding-at*, or more formally, in terms of (PT') rather than (PT). The system validates principles (EC), (TC) and (a'), but it fails to validate (b')-(e'), and thus falls short as a faithful formalization of Fabri's logic of composite modals. However, this problem can easily be solved. In order to 'regain' principles (b')-(e'), we need to have (ERed) available (because then the proofs, as laid out in Appendix II, will go through again). Recall that in *ML*, (ERed) can be derived from (PT), and is thus available as a theorem. By contrast, in ML^0 we have weakened (PT) to (PT'), and therefore (ERed) is no longer derivable. However, there is nothing to stop us from adding (ERed) after all, no longer as a derived theorem, but rather as an independent assumption. The resulting system will be called *ML'*. Once again, this system has precisely the same language $\mathcal{L}$ and precisely the same semantic clauses as *ML* and ML^0. Furthermore, it once again works with models $\mathbb{M} = \langle W, R, V, ex \rangle$. However, *ML'* uses (PT') to characterize possible times (just like ML^0 and unlike *ML*). Finally, *ML'* assumes (ERed) as an axiom, i.e. $\mathbb{M}, t \vDash E\varphi$ for all propositions φ, or semantically speaking, $ex(t) = \mathcal{L}$ (recall fn. 9 on p. 46), for all models $\mathbb{M}$ and for all times t.

It is easy to check that *ML'* is a strictly stronger semantics than ML^0, and hence continues to validate principles (EC), (TC) and (a'). Furthermore, because of the additional assumption (ERed), *ML'* also validates principles (b')-(e') (cf. the proofs in Appendix II). We thus find that *ML'* is a faithful representation of Fabri's account of composite modals, just like *ML*. As was already said earlier in this chapter, we presently do not have a definitive argument for either *ML* or *ML'* as the unique best formal reconstruction of Fabri. Furthermore, it is important to realize that despite their similar stances vis-à-vis Fabri, the systems *ML* and *ML'* come from two very different philosophical perspectives on the notion of possible time, viz., in terms of *holding-at* (cf. principle (PT) in the system *ML*) or in terms of *truth* (cf. principle (PT') in the system *ML'*).

The point is that since Fabri remains unclear on many aspects of his logic of composite modals, several model-theoretic reconstructions are conceivable that all seem to be equally appropriate representations of it; and as they have mutually different properties, they will lead to rather different conclusions about Fabri's original system. It seems, however, that all reconstructions will have to agree on at least one point. They will all have to somehow make sure that the propositional existence condition (ERed) is satisfied across times, whether it be as an axiom or as a theorem, for this is a *sine qua non* to prove (a')-(e'). Thus, *any* reconstruction would appear to elicit the conclusion that Fabri altogether avoids drawing in the issues relating to propositional existence. His logic, it seems, does not push the token-based approach to its limits, but integrates a moderate version of it instead. This observation is solidified as statement $\mathbf{A_8}$, which concludes this chapter.

- $\mathbf{A_8}$: The study of a number of implications that involve composite modals suggests that Fabri works under the assumption that propositional existence is satisfied by definition, and thus does not push the token-based approach to its limits.

References

Arlig A (2019) Medieval mereology. In: Zalta EN (ed) The Stanford encyclopedia of philosophy (Fall 2019 edition). https://plato.stanford.edu/archives/fall2019/entries/mereology-medieval/. Accessed 3 Dec 2020

Copeland J (2002) The genesis of possible world semantics. J Philos Log 31:99–137

Dutilh Novaes C (2005) Buridan's *consequentia*. Consequence and inference within a token-based semantics. Hist Philos Logic 26:277–297

Gabbay DM, Woods J (eds) (2006) Handbook of the history of logic, vol. 7: Logic and the modalities in the twentieth century. North-Holland, Amsterdam

Henry DP (1991) Medieval mereology. B.R. Grüner, Amsterdam

Hughes GE, Cresswell MJ (1996) A new introduction to modal logic. Routledge, London

Read S (2020a) Swyneshed, Aristotle and the rule of contradictory pairs. Log Univers 14:27–50

Read S (2020b) The rule of contradictory pairs, insolubles and validity. Vivarium 58:275–304

Whitaker CWA (1996) Aristotle's *De interpretatione*. Contradiction and dialectic. Clarendon Press, Oxford

Chapter 5
Fabri's Logic of Composite Modals in its Historical Context

Abstract This chapter situates Fabri's logic of composite modals in its historical context. Building on the results from Chap. 3, we focus our attention on the two most remarkable features of this logic: the allowance for quantification over *dicta* and quantification invariance. We argue that these features connect Fabri with authors from the Paris circle of John Mair, and that both features can be traced back to the work of John Buridan and his associates. The main conclusion of this chapter aligns with the one of Chap. 2: Fabri's logic of composite modals shares deep similarities with logics of authors who wrote in the modernist tradition.

Keywords John Fabri of Valenciennes · Composite modals · Quantification invariance · John Mair · John Buridan · *Via moderna*

In Chap. 3 and 4, we outlined some of the main features of Fabri's logic of composite modals, and we also suggested a model-theoretic reconstruction of the temporal framework that Fabri presupposes but fails to make explicit. In this chapter, we will situate Fabri's logic in its historical context. The guiding thread of our discussion will be two core aspects of Fabri's account: the allowance for quantification over *dicta* and what we called 'quantification invariance', i.e. the idea that quantifying over *dicta* is truth-conditionally irrelevant. Note that our goal in this chapter is to identify those authors with whom Fabri shares some of his views on how a logic for composite modals should look like. That is, our focus will be on what Fabri has *in common* with other authors, rather than on what is *distinctive of* his own thought. This approach appears to be justified. As we pointed out in Chap. 1, it seems reasonable that we can investigate the originality of an author only if we have a very clear and distinct idea of the logical landscape in which he worked, and given the present state of research into post-medieval logic, our idea of the backdrop against which Fabri should be understood, is rather murky to say the least. Indeed, as this chapter will set out some of the debated issues in modal logic during the post-medieval period, it will (hopefully) contribute to an improved understanding of the nature of post-medieval modal logic generally.

The idea that *dicta* bear quantification and quantification invariance are not principles that the average post-medieval author writing on modal logic accepts.

C. Geudens, L. Demey, *The Modal Logic of John Fabri of Valenciennes (c. 1500)*, SpringerBriefs in Philosophy, https://doi.org/10.1007/978-3-030-98802-9_5

Most authors believe that *dicta* resist being quantified over. Of all templates from scheme (*b*) they would only accept the ones in scheme (*c*) as being well-formed.

$$(c)\quad \text{That}\left[\begin{array}{l}\text{every}\\ \text{some}\\ [\textit{indefinite}]\\ \text{this}\end{array}\right. A \left\{\begin{array}{l}\text{is}\\ \text{is not}\end{array}\right. B \left\{\begin{array}{l}\text{is}\\ \text{is not}\end{array}\right. \left[\begin{array}{l}\text{necessary}\\ \text{possible}\\ \text{contingent}\\ \text{impossible}\end{array}\right.$$

Moreover, while Fabri considers the templates in (*c*) as *indefinite*, the mainstream view holds that they are *singular*. Most authors maintain that the *dictum* in composite modals functions as a singular term. It is therefore just as erroneous to quantify over *dicta* as it is to quantify over proper names like 'Socrates' or 'Plato'. For instance, Versoris writes that the quantity of a composite modal φ can be determined either with regard to φ as a propositional unit, or with regard to the *dictum* in φ in specific. In the latter case, the quantity of φ reduces to the quantity of its *dictum*, and so φ is either universal, particular, indefinite, or singular. In the former case, the quantity of φ is invariable, and φ is singular by definition:

> [*f*] It is asked whether every composite modal proposition is singular. It is replied that just as there is a twofold composition in such propositions, there is also a twofold subject, and thus the quantity can be considered in two ways. The first composition is that of the parts of the *dictum*, whose subject term is called the 'subject of the expression'. And if the quantity is determined with regard to this subject, then all composite modals have the same quantity as the underlying propositions of their *dicta* ... The second composition is that of the entire modal proposition, whose subject term is the entire *dictum*. And if the quantity is determined with regard to this subject, then all composite modals are singular.[1]

Versoris' account is echoed in Lambert of 's Heerenberg, among others[2]; and authors such as John of Caulincourt (or de Magistris; fl. c. 1480), Gerard of Harderwyck (†1503), Nicholas Tinctoris (†1498), and John Heynlin (de Lapide; c. 1430–1496), too, state that *dicta* in composite modals function as singular terms and thus resist being quantified over.[3] As Peter Crockaert (†1514) points out, *dicta* supposit materially for only one underlying assertoric proposition on this account, to

---

[1] Versoris, *in De int.* (Versoris 1497: f. 69ra): "Dubitatur an omnis propositio modalis sit singularis. Respondetur quod sicut duplex est compositio in propositione modali, ita potest esse duplex subiectum, et ita dupliciter potest attendi quantitas. Est enim ibi compositio partium dicti cuius subiectum locutionis subiectum vocatur. Et attendendo quantitatem ex parte istius subiecti modales sunt eiusdem quantitatis sicut et sue propositiones preiacentes ... Alia est compositio totius enunciationis modalis cuius enunciationis subiectum est totum dictum. Et attendendo quantitatem ex parte istius subiecti omnes propositiones modales singulares sunt."

[2] De Monte, *in Sum. log.* (de Monte 1489: ff. 30vb, 32vb).

[3] De Harderwyck, *in Sum. log.* (de Harderwyck 1488: sig. F $vi^{rb\text{-}va}$); de Lapide, *in De int.* (de Lapide [1495?]: sig. m 7ra); de Magistris, *in De int.* (de Magistris 1490: sig. h 4vb); Tinctoris, *in Sum. log.* (Tinctoris 1486: sig. F 5ra).

the effect that in 'That Sortes is running is possible' (*Sortem currere est possibile*) the *dictum* 'does not supposit univocally for every proposition 'Sortes is running', but only for one, for the meaning [of the modal proposition] is that it is possible that Sortes is running, that is, that this proposition 'Sortes is running' is possible.'[4] Most authors do not believe that *dicta* are suitable structures to bear quantification, and, consequently, the very concept of quantification invariance is usually not even considered.

The concept of modality occuring in the average post-medieval logic of composite modals is a clear anticipation of Quine's first grade of modal involvement, according to which modalities are (metalinguistic) statement-forming predicates attaching to the (metalinguistic) names of statements.[5] This mainstream account, which, as the above examples already suggested, is endorsed by virtually all traditionalists, is not an innovation of post-medieval modal logic, however. It regularly occurs in the sources from at least the twelfth century onwards. For instance, already in the *Dialectica Monacensis*, an anonymous text from around 1200 that is usually classified among the so-called 'proto-terminist' treatises, we find an account that is very similar to Versoris': a composite modal, its author says, can be of any quantity if we lay down that its quantity is identical with the quantity of its *dictum*; but if we take the quantity of a composite modal to be determined by the nature of the composition of its *dictum* and its modality, then all composite modals are singular.[6] The idea that composite modals as propositional units are invariably singular occurs in several authors from the thirteenth century, including Robert Kilwardby (c. 1215–1279), William of Sherwood (c. 1205-c. 1270), Roger Bacon (c. 1215–1292), Lambert of Lagny (fl. c. 1250), Aquinas (*De propositionibus modalibus*), and pseudo-Aquinas (*Summa totius logicae*).[7] For the fourteenth century it is attested in Richard Campsall (c. 1280–1350), Ockham, and Burley, among others.[8] Thus, most post-medieval authors adopted an approach that had been around for at least four centuries, and that, under a different guise, is still being debated in present-day (philosophy of) modal logic.[9]

[4] Crockaert, *in De int.* (Crockaert 1514: f. $60^{rb\text{-}va}$): "... dictum 'Sortem currere' non supponit vniuoce pro qualibet tali propositione 'Sortes currit', sed pro vna tantum, quia sensus est Sortem currere est possibile, id est, hec propositio 'Sortes currit' est possibilis."

[5] Quine (1976: 156–174).

[6] [Anon.], *Dialectica Monacensis* (de Rijk 1967-ii: 480 [ll. 7–15]). For an elaborate discussion of views of the matter in the age of Abelard, see Binini (2022: 45–74).

[7] Bacon, *Summulae dialectices*, II.1.6 (De Libera 1986: 257); De Latiniaco, *Logica*, I (Alessio 1971: 30). For Sherwood, Kilwardby, Aquinas and pseudo-Aquinas, see the comments in Jacobi (1980: 128); Thom (2019: 107); Uckelman (2008: 394).

[8] Lagerlund (2000: 61–62, 99).

[9] See esp. Stern (2016) for a discussion of the predicate approach to modality in contemporary modal logic.

The post-medieval authors who endorse an account like the one that we find in Fabri either are self-proclaimed *moderni* or are known to have been influenced by the *via moderna*. Among them are the anonymous *Compilatores* of the Paris Collège de Navarre, Dullaert, Nicholas of Amsterdam, George of Brussels, Tartaret, Pardo, Enzinas, Celaya, Mair, and Caubraith. All these authors maintain that *dicta* in composite modals are common terms rather than proper names, and they consequently also maintain that such *dicta* can be quantified over. Like Fabri but unlike most traditionalists, they hold that the templates in scheme (*b*) are all well formed, and that the ones in (*c*) are indefinite rather than singular. Moreover, they all agree that *dicta* in composite modals have material supposition, and they restrict the set of *supposita* of such a *dictum* to the *dictum*'s underlying assertoric propositions that exist at the moment of utterance.[10] The passages in quotes [*g*] and [*h*], which are resp. taken from Pardo's *Medulla dyalectices* and the *Compilatores*' commentary on Peter of Spain, provide illustration.

[*g*] ... The proposition 'that a man is running is possible' means that such a proposition 'man is running' is possible. And composite modals can be universal, particular, indefinite and singular ... Supposing there to be material supposition [of the *dictum*], as follows: 'Every that a man is running is possible' is universal, and it means that every such proposition 'man is running' is possible. The proposition 'some that a man is running is possible' is particular, and it

[*h*] If the *dictum* is taken materially, then it supposits for a proposition that consists of the subject and predicate terms of the *dictum*, but in the nominative case, where the copula is the copula of the *dictum* with the infinitive mood changed into a finite mood according to the exigencies of tense, like 'that Sortes is running' supposits for 'Sortes is running'; and such a *dictum* does not supposit for one single proposition, but

[10] Bruxellensis, *in Sum. log.* (Bruxellensis 1515: f. 16va); Caubraith, *Quadrupertitum* (Caubraith 1516: ff. 122va, 123rb); [Compilatores], *in Sum. log.* ([Compilatores] [1495]: sig. d vi^{va}); de Celaya, *in Sum. log.* (de Celaya 1525: sig. M vi^{ra}); de Amsterdammis, *in De int.* (Bos 2016: 339 [ll. 16–22]); Dullaert, *in De int.* (Dullaert 1515: f. 123ra); Enzinas, *in Sum. log.* (Enzinas 1528: f. 33ra); Major, *Introductorium perutile* (Major 1527: f. 67ra); Pardo, *Medulla dyalectices* (Pardo 1505: ff. 106va-107ra); Tartaretus, *in De int.* (Tartaretus 1503: ff. 56rb, 57rb), *in Sum. log.* (Tartaretus 1498: f. 15va). See Coombs (1990: 64–72) for further discussion.

means that some such proposition 'man is running' is possible. The proposition 'that a man is running is possible' is indefinite, and 'this that a man is running is possible', pointing at the proposition 'man is running' that is uttered by Socrates, is singular.[11]

rather for any proposition that is similar to it [i.e., the *dictum*].[12]

Notice that in quote [*h*] it is not explicitly asserted that *dicta* supposit materially in composite modals. Earlier in the discussion, however, the *Compilatores* pointed out that one of the idiosyncrasies of composite modals is precisely that the *dictum* always has material supposition.[13] Thus, the point of quote [*h*] is that the *dictum* in composite modals supposits materially for *all* existing underlying propositions rather than *one single* existing underlying proposition. (It is common currency in texts from this period to describe the relation between a *dictum* and its underlying assertoric propositions as a relation of similarity or correspondence; see e.g. quote [*i*] below.)

As we have seen, the principle that *dicta* in composite modals are common terms that supposit materially, along with the further specification that the sets of *supposita* of such *dicta* are restricted to occurrences of the same proposition type, jointly entail quantification invariance. This principle is thus valid in the modal logics of the *moderni* just mentioned. Most of them do not discuss the truth conditions of composite modals in much detail, however, and none of them expressly commits to quantification invariance. But there is evidence that some of them – Caubraith, Enzinas, and Pardo – were aware that the principle holds in their modal logics. As with Fabri, the proof is provided by their accounts of how we should determine the

[11] Pardo, *Medulla dyalectices* (Pardo 1505: f. 107ra): "... sensus istius 'hominem currere est possibile' est iste: talis propositio 'homo currit' est possibilis. Et possunt fieri vniuersales, particulares, indiffinite et singulares ... supponendo esse acceptionem materialem, hoc pacto. 'Omne ly hominem currere est possibile' est vniuersalis, et sensus est: omnis talis propositio 'homo currit' est possibilis. Ista est particularis: 'quoddam ly hominem currere est possibile'; et sensus est: quedam talis propositio 'homo currit' est possibilis. Et ista indiffinita: 'hominem currere est possibile'. Ista vero singularis: 'hoc ly hominem currere est possibile', demonstrando istam propositionem 'homo currit' a Sorte prolatam."

[12] [Compilatores], *in Sum. log.* ([Compilatores] [1495]: sig. d vi^{va}): "Si capiatur materialiter tunc sumitur pro propositione composita ex subiecto et predicato dicti in nominatiuo et copula infinitiui modi mutata in copulam [copulatiuam ed.] modi finiti secundum exigentiam temporis, vt 'Sortem currere' tantum valet sicut 'Sortes currit'. Et tale dictum non solum supponit pro vna propositione, sed pro qualibet sibi simili."

[13] See [Compilatores], *in Sum. log.* ([Compilatores] [1495]: sig. d vi^{va}): "Tertio <modalis diuisa et modalis composita> differunt quia in compositis vnum extremorum supponit materialiter [sc. the *dictum*], et in diuisa non oportet." Notice that the *Compilatores* maintain that the dictum is the *only* categorematic term in composite modals that has material supposition. Like Fabri, they appear to endorse the view that the modality has personal supposition.

truth value of composite modals.[14] They all state that the truth or falsity of such propositions can be assessed *per officiationem*, a procedure that is similar in nature to Fabri's analysis *per positionem inesse*: pick an arbitrary *suppositum* of the *dictum* – a so-called *propositio officians* –, and determine whether that *suppositum* is among the *supposita* of the modality.[15] The *officiatio* procedure as described in these texts does not differentiate between propositional quantities, and will thus assign any composite modal the same truth value as its co-existing quantifier mates (with 'quantifier mate' as defined in Sect. 3.3). Just as *positio inesse*, the procedure seems to have been developed starting from the observation that composite modals are invariant under quantification. One of the most informative discussions of *officiatio* is Pardo's:

> [*i*] [I]f the meaning of the composite modal, as it is explicated with the aid of a proposition corresponding with the *dictum*, is true, then the composite modal itself is true; but if the meaning is false, then the composite modal is false. This rule appears to be grounded in the fact that such a *dictum* supposits for propositions that correspond with itself, and thus one is allowed to replace the *dictum* by a proposition that corresponds with it. For instance, in 'That Sortes is running is possible', 'that Sortes is running' supposits for [occurrences of] the proposition 'Sortes is running', as the common view among logicians has it . . ., and thus the meaning of 'That Sortes is running is possible' is appropriately stated as follows: such a proposition 'Sortes is running' is possible. And by this meaning the truth and falsity of composite modals can be known. This is what some people mean when they say that this proposition 'that Sortes is running is possible' and other composite modal propositions are proven by *officiatio* . . .[16]

These modernist authors – and Fabri with them – are all drawing, either directly or indirectly, on Buridan and his associates. Buridan elaborates on the syntax and truth conditions of composite modals in several of his works, most notably in the *Summulae de propositionibus*, which forms part of the *Summulae de dialectica*,

---

[14]Caubraith, *Quadrupertitum* (Caubraith 1516: f. 123vb); Enzinas, *in Sum. log.* (Enzinas 1528: f. 33ra); Pardo, *Medulla dyalectices* (Pardo 1505: f. 106$^{va\text{-}b}$).

[15]The *officiatio* procedure was developed in treatises on the so-called 'proofs of terms and propositions' (*probatio terminorum, probatio propositionum*), which constituted a logical genre that flourished during the fourteenth and fifteenth centuries, and seems to have been particularly popular in the British Isles. Spade (2000) notoriously included the *probationes propositionum* among the key components of medieval logic that remain utterly unclear to modern scholars, and it seems that not much progress has been made on the topic over the course of the past two decades. Some important primary texts have been edited by De Rijk (1982). For more on *officiatio* in late-medieval modal logic, see Coombs (1990: 60–63).

[16]Pardo, *Medulla dyalectices* (Pardo 1505: f. 106va): "[S]i sensus modalis composite explicatus per propositionem dicto correspondentem sit verus, ipsa est vera, si falsus, ipsa est falsa. Ista regula in hoc videtur fundamentum accipere, quia tale dictum supponit pro propositione sibi correspondente, ergo loco illius dicti licet ponere propositionem tali dicto correspondentem, vt dicendo 'Sortem currere est possibile', ly Sortem currere supponit pro illa propositione 'Sortes currit', vt communis logicorum est sententia . . . ergo conuenienter datur sensus: talis propositio 'Sortes currit' est possibilis. Et ita per eius sensum habet veritas et falsitas cognosci. Hoc est quod aliqui dicunt quod ista propositio 'Sortem currere est possibile' et alie modales composite probantur officialiter. . ."

the *Questiones longe* on *De interpretatione*, and the *Tractatus de consequentiis*.[17] Differences in emphasis and detail aside, the accounts presented in these works are identical. On all three occasions, Buridan states that *dicta* in composite modals are common terms that can be quantified over and supposit materially for their underlying assertoric propositions existing at the moment of utterance. Unlike many of his post-medieval epigones, Buridan explicitly acknowledges that he is thus committed to quantification invariance. In the ninth conclusion to book II of the *Tractatus de consequentiis*, he writes:

> [*j*] In all composite modals in which the *dictum* is the subject term, from a particular there follows a universal, the rest being unchanged. For example, this follows: 'Some proposition '*B* is *A*' is possible, so every proposition '*B* is *A*' is possible', and similarly for truth and falsity, contingency and necessity. The reason is that among all the propositions '*B* is *A*', each signifies whatever the others signify and altogether as the others signify. So if things are as one signifies, then they are as any other signifies, and if things are not as one signifies, then they are not as any other signifies; so if one is true, then the other is true, and if one is false, then the other is false. Similarly for possibility, necessity and other modes. What I say concerning particular to universal, so I say concerning singular to universal. For this proposition, namely, 'The proposition '*B* is *A*' is possible' cannot be true unless 'Every proposition '*B* is *A*' is possible' is true.[18]

Buridan's position was taken over by Albert of Saxony, who even gives an almost verbatim quote of passage [*j*] in the *Perutilis logica*.[19] The gist of Buridan's account also occurs in the commentary on the *Analytica priora* of Marsilius of Inghen, as well as the commentary on the same work of the Pseudo-Scot, writing around the same time as Buridan.[20] The Pseudo-Scot, too, maintains that 'the *dictum* [in composite modals] should be taken materially, namely as suppositing for [occurrences of] the proposition that corresponds to it', where just as in quotes [*h*] and [*i*] the correspondence relation in question is the relation between the *dictum* and its underlying assertoric propositions. He states that composite modals can be of universal, particular or singular quantity, yet adds

---

[17] Buridanus, *in De int.* II, q. 7 (van der Lecq 1983: 76–79), *Summulae de dialectica* I.8.9.3 (van der Lecq 2005: 108–112), *Tractatus de consequentiis* II.7 (Hubien 1976: 69–78). For reconstructions of Buridan's modal logic, see esp. Hodges and Johnston (2017); Hughes (1989); Johnston (2015). All three studies focus on Buridan's logic for divided modals, however. His logic of composite modals has not yet been explored in depth. Some information is found in Lagerlund (2000: 143–149) and Thom (2003: 169–191).

[18] Buridanus, *Tractatus de consequentiis* II.7 (Hubien 1976: 70–71 [ll. 42–55]). Translation by Read (2015: 106), with modifications. See Read (2015: 35) for further comments on this passage, *pace* Lagerlund (2000: 137). Buridan also briefly comments on quantification invariance at Buridanus, *in De int.* II, q. 10 (van der Lecq 1983: 96 [ll. 8–17]).

[19] De Saxonia, *Perutilis logica* III.5, IV.6 (Berger 2010: 458–460 [ll. 6–24, 1–14], 706–708 [ll. 3–25, 1–6]). See Ely (1981: 54–55); Lagerlund (2000: 185–191) for further discussion.

[20] De Inghen, *in An. pr.* I (de Inghen 1516: f. 12[ra]); pseudo-Scotus, *in An. pr.* I (Wadding 1639: 310[a]). On the dating of the Pseudo-Scot's commentary on the *Analytica priora*, see Read (2015: 4–5).

that their quantity is mostly (*ut in pluribus*) indefinite.[21] Since the Pseudo-Scot remarks, with regard to the supposition of the *dictum* 'That every man is an animal' (*omnem hominem esse animal*), that 'the same judgment applies to this proposition 'Every man is an animal' as to every other proposition similar in kind with regard to truth, falsity, necessity and possibility', he, too, shows awareness that he is committed to quantification invariance.[22]

John Dorp gives an account that proceeds along the same lines, although as one of the few fourteenth-century Buridanians he refrains from accepting quantification invariance due to doubts regarding the material supposition of the *dictum*.[23] Dorp states that the material *supposita* of the *dictum* certainly *include* the underlying assertoric propositions of the *dictum*, but he is unsure whether its *supposita* are *restricted* to these propositions, which we saw is a necessary condition for quantification invariance to follow. He refers to the position taken by 'some people' (*aliqui*) – which authors Dorp has in mind is unclear – according to which the *dictum*'s material *supposita* also include occurrences of the *dictum* itself. On this interpretation, universal-affirmative composite modals are false by definition, since there are *supposita* of the *dictum* of which the modality cannot be verified, viz. the occurrences of the *dictum* itself – the *dictum* is a term and is thus not even the right kind of structure to occur among the *supposita* of the modality, which are all propositions. Now quantification invariance does not hold, since the entailment from non-universal composite modals to their universal quantifier mates is invalid. The relevant passage is as follows:

> [*k*] It is doubted whether a composite modal with a particularly quantified *dictum* and a composite modal with a universally quantified *dictum* are equivalent to each other ... To this doubt some people respond that they are equivalent, which they prove as follows. The proposition 'Some that an animal is a substance is possible' means that some such proposition 'An animal is a substance' is possible; and the proposition 'Every that an animal is a substance is possible' means that every such proposition 'An animal is a substance' is possible; but if one such proposition is possible, then every such proposition is possible. But this reply does not appeal to some people. For it is possible that a composite modal with a particular *dictum* is true while its corresponding composite modal with a universal *dictum* is false. For the *dictum* in a composite modal supposits materially not only for a proposition, but also for itself, and thus this proposition is true: 'Some that a man is an animal is possible', for it is a particular-affirmative proposition whose predicate agrees with the subject with regard to some *suppositum*, namely this proposition 'A man is an animal'. But this proposition is false: 'Every that a man is an animal is possible', for this subject not only supposits for this proposition but also for itself; but such a *dictum* is not possible since it is

[21] Pseudo-Scotus, *in An. pr.* I (Wadding 1639: 310[a]): "... semper sumendum est dictum materialiter in propositione, scilicet, prout supponit propositione sibi correspondente ... Tertio notandum quod vt in pluribus, propositiones de modo in sensu composito sunt indefinitae, vt verbi gratia ista est indefinita 'Possibile est Socratem currere', quia sua vniuersalis est ista: 'Omne possibile est Socratem currere' ..."

[22] Pseudo-Scotus, *in An. pr.* I (Wadding 1639: 310[a]): "... de ista propositione 'Omnis homo est animal' et de omni sibi consimili est idem iudicium quantum ad veritatem, falsitatem, necessitatem et possibilitatem."

[23] Dorp, *Perutile compendium* (Dorp 1499: sig. d 5[vb]-6[ra]).

> not a proposition. From this it follows that this proposition with a universal *dictum* is false, since it is a universal-affirmative proposition whose subject term supposits for something for which its predicate term does not supposit.[24]

Dorp's commentary on Buridan was a reference work for post-medieval *moderni*, being printed a handful of times around the turn of the sixteenth century. Even though Dorp does not appear to endorse quantification invariance himself, it leaves little doubt that he played an important role in transmitting the Buridanian approach to composite modals to late-medieval authors.

The question of whether *dicta* and modalities are appropriate structures for quantification has a long history, and Jacobi has shown that it was already being debated in the twelfth century, shortly after the very concept of *dictum* was introduced.[25] Buridan and his associates, however, were the first group of authors to explore the issue in depth, and the many references to especially Buridan by *Wegestreit moderni* leave no doubt that the texts from Buridan's circle really are the sources of the post-medieval accounts. The conclusion thus should be that Fabri's logic of composite modals (1) represents the modernist approach towards the topic, and (2) is ultimately rooted in the work of John Buridan and Albert of Saxony.

References

Alessio F (ed) (1971) Logica. Summa Lamberti. La nuova Italia, Firenze

Berger H (ed, trans) (2010) Albert von Sachsen: *Logik*. Felix Meiner, Hamburg

Binini I (2022) Possibility and necessity in the time of Peter Abelard. Brill, Leiden

Bos EP (ed) (1985) John of Holland. Four tracts on logic *(Suppositiones, Fallacie, Obligationes, Insolubilia)*. Ingenium, Nijmegen

---

[24] Dorp, *Perutile compendium* (Dorp 1499: sig. d 6ra): "Dubitatur vtrum modalis composita de dicto particularisato et modalis composita de dicto vniuersalisato equiualeant ... Ad istud dubium respondent aliqui quod sic et declarant. Nam sensus istius 'Aliquod animal esse substantiam est possibile' est iste: aliqua talis propositio 'Animal est substantia' est possibilis. Et sensus huius 'Omne animal esse substantiam est possibile' est iste: omnis talis propositio 'Animal est substantia' est possibilis, modo si aliqua talis sit possibilis et quelibet talis est possibilis. Sed ista responsio aliquibus non placet, nam stat modalem compositam de dicto particularisato esse veram modali composita de dicto vniuersalisato existente falsa. Nam dictum in modali composita supponit materialiter non solum pro propositione sed etiam pro semetipso, et sic ista est vera 'Aliquod hominem esse animal est possibile', quia est vna particularis affirmatiua cuius predicatum conuenit subiecto pro aliquo eius supposito, scilicet pro ista 'Homo est animal'; sed ista est falsa 'Omne [m] hominem esse animal est possibile', quia illud subiectum non solum supponit pro illa propositione sed etiam pro semetipso, modo tale dictum non est possibile ex quo non est propositio. Quare sequitur quod illa de dicto vniuersalisato est falsa, quia est vniuersalis affirmatiua cuius subiectum supponit pro aliquo pro quo non supponit predicatum." The view that the *dictum* in composite modals supposits for occurrences of itself is also mentioned in the *Fallacie* of the Prague author John of Holland (fl. c. 1365), but John does not appear to endorse it; see de Hollandia, *Fallacie* (Bos 1985: 49–50).

[25] Jacobi (1980: 114).

Bos EP (ed) (2016) Nicholas of Amsterdam: commentary on the *Old Logic*. John Benjamins, Amsterdam

Bruxellensis G (1515) Interpretatio in Summulas magistri Petri Hyspani una cum magistri Thome Bricot questionibus de nouo in cuiusuis fine tractatus additis. Per Johannem de la Place, Lugduni

Caubraith R (1516) Quadrupertitum in oppositiones, conuersiones, hypotheticas et modales. In aedibus Iodoci Badii et Emundi Fabri, Parrhisiis

[Compilatores] [1495] Compilatio ex Buridano, Dorp, Ockan et aliis nominalibus in textum Petri Hyspani edita in regali collegio Nauerre Parisius. [Nicolaus Wolff], [Lyon]

Coombs JS (1990) The truth and falsity of modal propositions in Renaissance nominalism. Unpublished PhD dissertation, University of Texas, Austin

Crockaert P (1514) Acutissime questiones et quidem perutiles in singulos Aristotelis logicales libros. Impensis sumptibusque Joannis Petit et Gaufredi de Marnef, Parisiis

De Celaya J (1525) Expositio in primum tractatum Summularum Petri Hispani. Venalia prostant in Clauso Brunello, Parrhisiis

De Harderwyck G (1488) Copulata Petri Hyspani secundum processum Burse Laurentii. [Henricus Quentell], [Köln]

De Inghen M (1516) Questiones super libros Priorum analyticorum. Impensis haeredum Scoti et sociorum, Venetiis

De Lapide J [1495?] Libri artis logice Porphyrii et Aristotelis. Per Ioannem de Amerbach, Basileae

De Libera A (1986) Les *Summulae dialectices* de Roger bacon I-II. *De termino, De enuntiatione*. Archives d'Histoire Doctrinale et Littéraire du Moyen Âge 53:139–289

De Magistris J (1490) Questiones subtiles et perutiles super totum cursum logice Porphyrii et philosophi. Iussu et impensis Octauiani Scoti, Venetiis

De Monte L (1489) Textus omnium tractatuum Petri Hispani etiam sincathegreumatum [sic] et paruorum logicalium cum copulatis secundum doctrinam diui Thome Aquinatis iuxta processum magistrorum Colonie in Bursa Montis regentium. [Henricus Quentell], [Köln]

De Rijk LM (ed) (1967) *Logica Modernorum*. A contribution to the history of early terminist logic, vol. 2: The origin and early development of the theory of supposition, 2 vols. Van Gorcum, Assen

De Rijk LM (ed) (1982) Some 14th century tracts on the *probationes terminorum* (Martin of Alnwick O.F.M., Richard Billingham, Edward Upton and others). Ingenium, Nijmegen

Dorp J (1499) Perutile compendium totius logice Ioannis Buridani cum preclarissima expositione. per Petrum Joannem de Quarengiis, Venetiis

Dullaert J (1515) Questiones super duos libros Peri hermenias Aristotelis. Impresse per Stephanum Baland, [Lyon]

Ely P (1981) The modal logic of Albert of Saxony. Unpublished PhD dissertation, University of Toronto, Toronto

Enzinas F (1528) Primus tractatus Summularum. Apud Reginaldum Chauldiere, Parisiis

Hodges W, Johnston S (2017) Medieval modalities and modern methods: Avicenna and Buridan. IfCoLog J Log Appl 4:1029–1073

Hubien H (ed) (1976) Iohannis Buridani Tractatus de consequentiis. Publications universitaires, Leuven

Hughes GE (1989) The modal logic of John Buridan. In: Corsi G, Mangione C, Mugnai M (eds) Atti del Convegno Internazionale di Storia della Logica: Le teorie delle modalità. CLUEB, Bologna, pp 93–111

Jacobi K (1980) Die Modalbegriffe in den logischen Schriften des Wilhelm von Shyreswood und in anderen Kompendien des 12. und 13. Jahrhunderts. Funktionsbestimmung und Gebrauch in der logischen Analyse. Brill, Leiden

Johnston S (2015) A formal reconstruction of Buridan's modal syllogism. Hist Philos Logic 36:2–17

Lagerlund H (2000) Modal syllogistics in the middle ages. Brill, Leiden

Major J (1527) Introductorium perutile in Aristotelicam dialecticen duos terminorum tractatus ac quinque libros summularum complectens. In aedibus Ioannis Parvi et Aegidii Gorimontii, [Paris]
Pardo H (1505) Medulla dyalectices omnes ferme grauiores difficultates logicas acutissime dissoluens. Per Guillermum Anabat, Parisius
Quine WV (1976) The ways of paradox and other essays. Harvard University Press, Cambridge
Read S (trans) (2015) John Buridan: *Treatise on consequences*. Fordham University Press, New York
Spade PV (2000) Why don't mediaeval logicians ever tell us what they're doing? or, what is this, a conspiracy? Available online at http://pvspade.com/Logic/docs/Conspiracy.pdf
Stern J (2016) Toward predicate approaches to modality. Springer, Dordrecht
Tartaretus P (1498) Expositio super Summulis Petri Hyspani. Expensis Jacobi Maillet, Lugduni
Tartaretus P (1503) Expositio super textu logices Aristotelis. Per Lazarum de Soardis, Venetiis
Thom P (2003) Medieval modal systems, problems and concepts. Ashgate, Aldershot
Thom P (2019) Robert Kilwardby's science of logic. A thirteenth-century intensional logic. Brill, Leiden
Tinctoris N (1486) Dicta super Summulas Petri Hyspani. Per Michahelem Gryff, Reutlingae
Uckelman SL (2008) Three 13th-century views of quantified modal logic. Adv Mod Log 7:389–406
Van der Lecq R (ed) (1983) Johannes Buridanus: *Questiones longe super librum Perihermeneias*. Ingenium, Nijmegen
Van der Lecq R (ed) (2005) Johannes Buridanus: *Summulae de propositionibus*. Brepols, Turnhout
Versoris J (1497) Questiones pulcerrime in ueterem artem Arestotelis. Per Henricum Quentell, Coloniae
Wadding L (ed) (1639) Ioannis Duns Scoti opera omnia, vol 1. Sumptibus Laurentii Durand, Lugduni

Chapter 6
Fabri's Logic of Divided Modals

Abstract In this chapter, we continue our exploration of Fabri's modal logic by turning towards his logic of divided modals. We first discuss Fabri's account of the truth conditions of such modals and sketch the main features of his theory of ampliation. We then turn to the logical geometry of divided modals, and we show that Fabri took the opposition and entailment relations between quantified divided modals to constitute a modal octagon.

Keywords Modal logic · Divided modals · Ampliation · Logical geometry · Modal square of opposition · Modal octagon

As we have seen in Chap. 3, one of the main puzzles in Fabri's logic of composite modals is the functioning of *dicta*. This is not the case in his logic of divided modals. *Dicta* do not serve a specific function in this latter logic. Remember that Fabri maintains that there are two ways of construing divided modals: one that does, and one that does not, involve the use of *dicta*; and that he also maintains that for each divided modal that contains a *dictum*, an equivalent proposition can be formed that consists of the same categoremes and modality, but does not involve a *dictum* (see Chap. 2). The point is that *dicta* are actually superfluous in Fabri's logic of divided modals. If Fabri had chosen to restrict the use of *dicta* to the realm of composite modals and not to admit such structures in his logic of divided modals, then this latter logic would have had the exact same expressiveness. Also, while the supposition of *dicta* is a thorny issue in Fabri's logic of composite modals, it is a straightforward matter in his logic of divided modals. Since *dicta* do not appear as categoremes in divided modals, they do not supposit in such propositions. The subject terms of *dicta* do supposit – Fabri points out

C. Geudens, L. Demey, *The Modal Logic of John Fabri of Valenciennes (c. 1500)*, SpringerBriefs in Philosophy, https://doi.org/10.1007/978-3-030-98802-9_6

that they typically have personal supposition[1] –, but the *dicta* themselves do not. The two principal puzzles of Fabri's logic of divided modals are the truth conditions of such propositions and the opposition relations holding between them. These are the main topics that are treated in this chapter. In the next chapter, we will place Fabri's account in its historical context.

There is no passage in the questions on *De int.* 12–13 or *An. pr.* I.3, 8–22 where Fabri engages in a detailed discussion of the truth conditions of divided modals, but we can reconstruct Fabri's thoughts on the issue without too much trouble. There is evidence that also in his logic for divided modals Fabri treats existential import as an implication, meaning the (non-)emptiness of the logical subject of a divided modal φ has something to say about φ's truth value: if φ's logical subject is empty, then φ is false if φ is affirmative, but true if negative. For instance, the proposition 'For Socrates it is possible not to be' (*Sortem possibile est non esse*), which we shall see is negative by Fabri's principles, is true if Socrates does not exist, Fabri says.[2] Moreover, Fabri does discuss how we should assess the truth value of divided modals in a series of insightful comments at the end of the questions on *De int.* 12-13.[3] The procedure he introduces, relies on the 'reducing-to-assertoric' principle (*positio inesse*), which is also the mechanism behind Fabri's procedure for determining the truth value of composite modals (see Sect. 3.3). If we combine the existential import requirement with the information that can be extracted from the working of this procedure, then we get a complete picture of what Fabri considers to be the truth conditions of divided modals. We do this in Sect. 6.1. With these truth conditions clarified, we will turn to the opposition relations holding between divided modals, in Sect. 6.2. The next couple of paragraphs are introductory in nature, and discuss some basic features of divided modals: their syntax, quality, and quantity.

The study of **S**'s questions on *De int.* 12-13 and *An. pr.* I.3, I.8-22 indicates that divided modals involving one of the modalities in $\mathfrak{M}$ are instances of one of the templates in (*d*), which are 128 in total $((4 \times 2 \times 4 \times 2) \times 2)$. For future reference, we assume that these templates, which constitute the core of Fabri's logic of divided modals, together form the set T_{DM}. This set is the analogue for divided modals of the set T_{CM}. The list on the left-hand side in (*d*) denotes the subset of T_{DM} consisting of those templates that are construed without a *dictum* and where the modality is an adverb. Analogously, the list on the right-hand side consists of those templates that contain a *dictum* and where the modality in each template is an adjective. For each pair of facing templates, instances of these templates that are obtained by uniform substitution are equivalent to each other.[4]

[1]**S**, ll. 130–135.

[2]**S**, ll. 147–150. By stating that existential import is determined by propositional quality rather than quantity (as is common in present-day quantificational logic) Fabri is merely voicing the mainstream view among medieval authors. For more details on existential import in medieval logic after the thirteenth century, see esp. Ashworth (1973); Church (1965); Klima (2001).

[3]**S**, ll. 435–455.

[4]To save space, all modalities in (*d*) have been abbreviated. Also, quantifiers in (*d*) are defined up to logical equivalence, like the ones in (*a*)–(*c*).

$$(d)\quad \left[\begin{array}{l}\text{Every}\\ \text{Some}\\ [\textit{Indefinite}]\\ \text{This}\end{array}\right. A \left[\begin{array}{l}\text{is}\\ \text{is not}\end{array}\right. \left[\begin{array}{l}N\text{-ly}\\ P\text{-ly}\\ C\text{-ly}\\ I\text{-ly}\end{array}\right. \left[\begin{array}{l}B\\ \text{not } B\end{array}\right. \;\Bigg|\; \text{For} \left[\begin{array}{l}\text{Every}\\ \text{Some}\\ [\textit{indefinite}]\\ \text{This}\end{array}\right. A,\ \text{it} \left[\begin{array}{l}\text{is}\\ \text{is not}\end{array}\right. \left[\begin{array}{l}N\\ P\\ C\\ I\end{array}\right. \left[\begin{array}{l}\text{to be}\\ \text{not to be}\end{array}\right. B$$

Each template in (*d*) is either *affirmative* or *negative*. A template is affirmative if it does not contain a negation, whether explicit ('not') or implicit (contained in 'impossible'), or, more generally, if it contains an even number of negations (e.g., 'not impossible'). Consequently, a template is negative if it contains an odd number of negations, whether explicit or implicit. Thus 'Every *A* is not necessarily not *B*' is affirmative, while 'Some *A* is not impossibly not *B*' is negative.[5]

The quantity of a divided modal can be considered from two perspectives: there is its propositional quantity, on the one hand, and its modal quantity, on the other hand. The *propositional* quantity of a divided modal is determined by the quantity of its logical subject rather than its grammatical subject. Thus, both 'Every *A* is possibly *B*' and 'For every *A* it is possible to be *B*' are universal in this respect. The *modal* quantity of a divided modal is determined by whether the modality occurring in it distributes over times. Propositions of necessity and impossibility are modally universal, since 'necessary' and 'impossible' both distribute over times; while propositions of possibility and contingency are modally particular, since 'possible' and 'contingent' do not distribute over times. Thus, 'Every *A* is possibly *B*' and 'For every *A* it is possible to be *B*' are both particular in this respect. Propositions that are both propositionally and modally universal or particular are said to be universal or particular absolutely (*simpliciter*). Otherwise they are said to be of mixed quantity (*mixte quantitatis*):

> [*I*] It should be noted, with regard to the quantity of [divided] modal propositions, that those propositions are singular with respect to the subject (*ex parte subiecti*) in which the subject is a singular term, as in 'For Socrates it is possible to be running', and those propositions are universal in which the subject is a common term that is determined by a universal sign. Likewise for particular and indefinite quantity. But with regard to the modality (*ex parte modi*), divided modal propositions of necessity and impossibility are universal, and propositions of possibility and contingency are particular. It is clear, therefore, that those propositions that are universal with respect to both the subject and the modality are universal absolutely (*simpliciter uniuersales*); and that those propositions that are particular with respect to both the subject and the modality are particular absolutely (*simpliciter particulares*). But those propositions that are universal with respect to one and particular with respect to the other can be said to be of mixed quantity (*mixte quantitatis*).[6]

Furthermore, it is worth noticing that none of the templates in (*d*) contains a sequence, or iteration, of modalities: each of them contains exactly one modality. The properties of iterated modalities in divided modals, such as 'necessarily

[5] See esp. **S**, ll. 144–169, 202–204, 269–276.

[6] **S**, ll. 24–32.

possibly' (*necessario possibiliter*), attracted the attention from several of Fabri's contemporaries, nearly all of them authors from Mair's circle in Paris.[7] There are no indications that Fabri considered propositions involving such sequences as well-formed, however.

6.1 The Truth Conditions of Divided Modals

We are now in a position to discuss Fabri's truth conditions of divided modals. Our focus in this section will be on divided modals that contain a *dictum*. The restriction is harmless, as the case of divided modals without a *dictum* is entirely analogous, and it allows for a considerable simplification of the discussion. Note that in what follows the term 'divided modal' means 'divided modal with a *dictum*', and that whatever is said about the quantity of divided modals should be understood in terms of *propositional* quantity, unless otherwise indicated.

Let us start with the case of singular divided modals. Following quote [*m*], such propositions are reduced-to-assertoric in the same way as composite modals (see Sect. 3.3). Let φ be a singular divided modal: φ is reduced-to-assertoric by construing a proposition ψ that is an underlying proposition of φ's *dictum*. The truth value of φ is determined by checking whether φ's modality $m \in \mathfrak{M}$ 'can be verified of' ψ, or, equivalently, whether ψ is among the *supposita* of *m*:

> [*m*] The second rule is that divided modals where the subject term is singular and non-connotative are reduced-to-assertoric in the way that has already been explained for composite modals. Thus 'For Socrates it is possible to be running' (*Sortem possibile est currere*) is reduced-to-assertoric by 'Socrates is running' (*Sortes currit*) ... And then we should see whether the modality occurring in the modal proposition is true of (*possit uerificari de*) this assertoric proposition. If so, then the modal proposition is true; if not, then it is false.[8]

Note that we should also check whether φ's logical subject has an existing referent, although Fabri does not say this explicitly in quote [*m*]. Thus, to stay with Fabri's own example, 'For Socrates it is possible to be running' should be labeled as true at time *t* by Fabri's principles just in case the term 'Socrates' is non-empty at *t* and 'Socrates is running' is among the *supposita* of 'possible' at *t*; that is, this proposition is true at *t* just in case Socrates exists at *t* and there is a *t'* such that 'Socrates is running' is true at *t'* if it exists at *t'*, or, less laboriously, such that 'Socrates is running' *holds* at *t'* (see Chap. 4 on the notion of holding-at). The truth conditions of singular divided modals of the form 'For this *A* it is *m* to be *B*', with $m \in \mathfrak{M}$, should thus be those in the table below. (The equivalences make use of the symbolism that

---

[7] The history of iterated modalities is not well known, but the study of the topic is presumably one of the innovations of post-medieval modal logic. Some information is found in Coombs (1990).

[8] **S**, ll. 435–443. The phrase 'as has already been explained' refers to quote [*e*], which immediately precedes the text in quote [*m*].

was introduced in Chap. 1. The symbols '$\ddagger 1$' and '$\ddagger 2$' indicate resp. the one- and two-sided reading of 'contingent'. Also note that divided modals of two-sided contingency where 'contingent' is not preceded by a negation sign are equivalent to conjunctions where one conjunct involves 'possible' and the other 'possible not'.[9] The conjunct involving 'possible' is affirmative, and thus true only if the subject term is non-empty. Since this conjunct has existential import, the entire conjunction has existential import, and thus the original proposition of two-sided contingency, with which the conjunction is equivalent, also has existential import.)

- $\underline{a} \mapsto B^*$, if expressed at time t, is true at t iff a is non-empty at t and for every t', if a proposition $\underline{a} \mapsto B$ exists at t', then it is true at t'
- $\underline{a} \mapsto B^\dagger$, if expressed at time t, is true at t iff a is non-empty at t and for some t', if a proposition $\underline{a} \mapsto B$ exists at t', then it is true at t'
- $a \mapsto B^\S$, if expressed at time t, is true at t iff either a is empty at t or for every t', if a proposition $\underline{a} \mapsto B$ exists at t', then it is not true at t'
- $\underline{a} \mapsto B^{\ddagger 1}$, if expressed at time t, is true at t iff $\underline{a} \mapsto B^\dagger$ is true at t
- $\underline{a} \mapsto B^{\ddagger 2}$, if expressed at time t, is true at t iff a is non-empty at t and for some t', t'', if a proposition $\underline{a} \mapsto B$ exists at t', then it is true at t', and if a proposition $\underline{a} \mapsto B$ exists at t'', then it is not true at t''

As for divided modals of particular, indefinite or universal quantity, Fabri maintains that reducing-to-assertoric is one and the same procedure in all three cases, and so we shall treat them together. If φ is a non-singular divided modal, then φ is reduced-to-assertoric by construing a *list* of assertoric propositions $\psi_1, \psi_2, \ldots, \psi_n$ from the *dictum* in φ, rather than one single assertoric proposition. Each ψ_i so construed (with $1 \leq i \leq n$) is a singular proposition with an indexical ('*hoc*', '*illud*') as its subject term and, as its predicate term, the predicate term of φ's *dictum* but in the nominative case. The indexical in every such ψ_i refers to a *suppositum* of φ's logical subject. For instance, in the case of (6) and (7) below, if A has n *supposita* at the moment of utterance, then applying *positio inesse* amounts to construing n propositions 'This is not B' (6) and 'This is B' (7), where the indexical in every such proposition picks out a different *suppositum* of A.

(6) For some A it is necessary not to be B
(7) For every A it is not necessary to be B

---

[9]See **S**, f. 241[rb]: "Sciendum quod propositio de contingenti in qua ante 'contingens' non ponitur negatio exponitur per unam copulatiuam ... ut 'Sortem contingit currere' exponitur per istam: 'Sortem possibile est currere et Sortem possibile est non currere'; et ista: 'Omnem hominem contingit currere' exponitur per istam: 'Omnem hominem possibile est currere et omnem hominem possibile est non currere'."

Fabri explains the procedure in the passage in quote [*n*].

> [*n*] The fourth rule is that a particular or indefinite divided modal is reduced-to-assertoric by many singular propositions in which [i] demonstrative pronouns pick out those things for which the subject in the modal proposition supposits (*ille res pro quibus supponebat subiectum in ipsa modali*). For instance, 'For something white it is possible to be black' (*album possibile est esse nigrum*) is reduced-to-assertoric as follows: 'This is black' (*hoc est nigrum*), 'That is black' (*illud est nigrum*), and so on, pointing successively to all *supposita* of 'white' (*consequenter demonstrando quodlibet pro quo supponebat 'album'*). [ii] And then we should see whether the modality occurring in the modal proposition accords with at least one such assertoric proposition (*competat alicui tali propositioni de inesse*). If so, then the modal proposition is true; if not, then it is false. The fifth rule is that universal divided modals are reduced-to-assertoric in the way that has been explained in the preceding rule. [iii] But after they have been reduced-to-assertoric, then we should see whether the modality occurring in the modal proposition can be truly predicated of every such assertoric proposition (*possit uere predicari de qualibet tali propositione de inesse*). If so, then the modal proposition is true; if not, then it is false.[10]

Clause [i] might suggest that the reference of the indexicals in the propositions ψ_i is tied to the present, with each indexical picking out a present *suppositum* of φ's logical subject. This is not the case, however. A widely accepted principle in modal logic from the thirteenth to the early sixteenth centuries is that the logical subject in non-singular divided modals is ampliated.[11] Fabri, too, accepts this principle. He does not mention it in quote [*n*] – or anywhere else in the commentary on *De int.* 12-13, for that matter –, but it does occur in the modal logic in his commentary on the *Analytica priora*, and several passages from the latter work prove that Fabri believes that non-singular divided modals where the modality is a member of $\mathfrak{M}$ ampliate their logical subject to supposit for both present and possible *supposita*.[12] Every divided modal φ where the logical subject *A* occurs ampliated is equivalent to an extended proposition where *A* has been substituted by the disjunctive term '× that is or can be *A*' (× *quod est vel potest esse A*), where × is a placeholder for either

[10] **S**, ll. 443–453.

[11] On ampliation, see the literature cited in fn. 10 on p. 14. Ockham was presumably the most notorious author not to endorse ampliation. On Ockham, see esp. Johnston (2015b); Priest and Read (1981).

[12] See, e.g., **S**, f. 231vb: "... uniuersales de impossibili in quibus non ponitur negatio post modum, si subiectum restringatur ad illa que sunt, non conuertuntur"; **S**, f. 239va: "... oportet subiecta propositionum de necessario ampliari ad supponendum pro his que possunt esse"; **S**, f. 240rb: "... bene sequitur 'omne currens possibile est non esse hominem, et omne rationale necessario est homo, ergo omne rationale necessario non est currens', quia sicut conclusio est falsa, ita et maior, saltem si ampliatur subiectum pro eo quod potest esse, qualiter debet capi." Epistemic modalities seem to ampliate their logical subject to supposit for both present, past, future, possible and imaginary *supposita*; see esp. **S**, f. 231vb-232ra, where Fabri notes that 'A chimera is believed to be a goat-stag' (*Chimera creditur esse hircoceruus*) converts into 'Something that is believed to be a goat-stag is, was, will be, can be, or is imagined to be a chimera' (*Aliquid quod creditur esse hircoceruus est, fuit, erit, potest uel imaginatur esse chimera*). The extension of ampliation to the imaginary, which is also briefly touched on at **S**, f. 194vb, was an innovation of Marsilius of Inghen, and it was a hotly debated issue in Fabri's day; see esp. Ashworth (1977); Ciola (2020).

'something' (*aliquid*, *quiddam*), if φ is particular or indefinite, or 'everything' (*omne*), if φ is universal. Since the actual *supposita* of A are among its possible *supposita*, the disjunction '$\times$ that is or can be A' can be simplified to its second disjunct '$\times$ that can be A'. For instance, (6) is equivalent to 'For something that is or can be A it is necessary not to be B', which can be simplified to 'For something that can be A it is necessary not to be B', while (7) is equivalent to 'For everything that is or can be A it is not necessary to be B', which can be simplified to 'For everything that can be A it is not necessary to be B'. That the subject term in such propositions φ is ampliated to the possible means, of course, that the reference of the indexicals in the propositions ψ_i should be interpreted accordingly. In each ψ_i, the indexical does not simply pick out something that is A, but rather something that is or can be A. In other words, the propositions ψ_i should not be defined on the set of actual *supposita* of φ's logical subject, but on its set of possible *supposita*.[13]

Following sentences [ii] and [iii] in quote [*n*], the basic idea for calculating the truth value of non-singular divided modals is this: if φ is particular or indefinite, then we check whether at least one ψ_i is among the *supposita* of φ's modality; while, if φ is universal, then we check whether all ψ_i are among the *supposita* of φ's modality. Again, we should check first whether φ's logical subject is empty, although this is not said explicitly in quote [*n*]. Thus, to determine the truth value of (6) at time t, we check whether A is empty at t, and, in case A is non-empty, whether at least one proposition 'This is not B' is among the *supposita* of 'necessary' at t. The latter condition is satisfied just in case at least one proposition 'This is not B' is true at t' if it exists at t', for every t'. Note that the reference of the indexical in each such proposition is rigid, or invariant across time: if 'This is not B' exists at a time t' different from t, then the indexical 'This' picks out precisely the same *suppositum* of A as it picks out at t. Thus, a proposition 'This is not B' is true at t' if it exists at t', for every t', just in case, for every t', either the *suppositum* that is picked out by the indexical at t does not exist at t', so that the reference of the indexical fails at t' (and the *negative* proposition 'This is not B' is vacuously true at t'), or the *suppositum* in question does exist yet is not among the *supposita* of B at t'. Likewise, to determine the truth value of (7) at t, we check whether A is empty at t, and, in case A is non-empty and has n *supposita* at t (with $n \geq 1$), we check whether all n propositions 'This is B' are excluded from the *supposita* of 'necessary' at t, which is the case iff for each proposition 'This is B', there is a time t' where it is false if it

[13] A reasonable question to ask at this point is: what *are* non-actual possible *supposita* according to Fabri? Are the non-actual possible *supposita* of a categoreme A at time t all and only conventional significates of A that actually exist at times t, t' and have property A at t' though not at t? Or do they also include those conventional significates of A that do not exist at t, exist at t' and have property A at t'? The answer is: we do not know. As pointed out, Fabri does not mention ampliation in the commentary on *De int.* 12-13, and also his remarks on the principle in the commentary on *An. pr.* I.3, 8-22 are rather cursory. None of the texts in **S** that we studied sheds light on this issue.

exists – be it due to reference failure of the indexical (in the *affirmative* proposition 'This is *B*') or due to the fact that the *suppositum* that is picked out by the indexical at *t* exists at *t'* yet is not among the *supposita* of *B* at *t'*. Fabri himself provides an example of how to analyze universal-affirmative divided modals in a parallel passage occurring in the commentary on *De int.* 3, 16b6-19, which further develops the main idea from sentence [iii] in quote [*n*]:

> [*o*] The modal proposition 'For every star it is possible to be seen by me' is reduced-to-assertoric by means of many singular propositions taken conjunctively: 'This star is seen by me', 'That star is seen by me' and so on, always pointing at a different star. Thus, in order for the modal proposition to be true it suffices that each of these propositions taken individually be possible.[14]

It is important to stress that for Fabri the truth values of divided modals eventually depend on the co-supposition of the names of propositions with modal predicates. Remember, the same goes for the truth value of composite modals as well as assertoric propositions (see Chap. 3). The point is that Fabri's suppositional semantics covers *both* assertoric *and* modal logic. The modal case involves more refined machinery than the assertoric case, certainly; but the core principle is the same in both cases.

Sentences [ii] and [iii] of quote [*n*] and the existential import requirement provide us with everything we need to reconstruct Fabri's truth conditions of non-singular divided modals. The conditions for divided modals of the form 'For $\times$ *A* it is *m* to be *B*' and 'For $\times$ *A* it is *m* not to be *B*', where $\times$ is a placeholder for either the indefinite, particular or the universal quantifier and $m \in \mathfrak{M}$, are listed in the table below. To save space, the one-sided contingent ($^{\ddagger 1}$) is not included, since propositions of one-sided contingency are equivalent to their corresponding propositions of possibility. Moreover, a possibility operator '†' attaching to a subject term *A* indicates that *A* is ampliated to the possible. This way of expressing ampliation reflects the fact that a string '$\times$ that is or can be *A*' can always be reduced to '$\times$ that can be *A*'. Propositions '$\underline{a}_i^{\dagger} \mapsto B$' and '$a_i^{\dagger} \mid B$' are formal renderings of the type of deictic propositions that are obtained from non-singular divided modals by the reducing-to-assertoric procedure. Here '$a_i^{\dagger}$' is a singular term picking out the *i*-th possible *suppositum* of *A* (where $1 \leq i \leq n$), assuming that *A* has *n* possible *supposita*. Thus, '$\underline{a}_i^{\dagger} \mapsto B$' is a translation of 'This is *B*', and '$a_i^{\dagger} \mid B$' of 'This is not *B*', where in both cases the indexical picks out the *i*-th possible *suppositum* of *A*.

[14] **S**, f. 199rb: "Ista modalis 'omnem stellam possibile est me uidere' ponitur inesse per istas multas singulares †coniunctim† sumptas: 'istam stellam uideo', 'istam stellam uideo', et sic de aliis, demonstrando semper aliam stellam. Unde satis est quod quelibet seorsum sit possibilis ad hoc quod illa modalis sit uera [possibilis **S**]."

- $\underline{A}^{\dagger} \mapsto B^{*}$, if expressed at time t, is true at t iff there are n *supposita* of A at t (with $n \geq 1$), and for every i, for every t', if $\underline{a}_i^{\dagger} \mapsto B$ exists at t', then it is true at t'
- $\underline{A}^{\dagger} \mapsto B^{\dagger}$, if expressed at time t, is true at t iff there are n *supposita* of A at t (with $n \geq 1$), and for every i, for some t', if $\underline{a}_i^{\dagger} \mapsto B$ exists at t', then it is true at t'
- $A^{\dagger} \mapsto B^{\$}$, if expressed at time t, is true at t iff (i) there are no *supposita* of A at t, or (ii) there are n *supposita* of A at t (with $n \geq 1$), and for every i, for every t', if $\underline{a}_i^{\dagger} \mapsto B$ exists at t', then it is not true at t'
- $\underline{A}^{\dagger} \mapsto B^{\ddagger 2}$, if expressed at time t, is true at t iff there are n *supposita* of A at t (with $n \geq 1$), and for every i, for some t', t'', if $\underline{a}_i^{\dagger} \mapsto B$ exists at t', then it is true at t', and if $\underline{a}_i^{\dagger} \mapsto B$ exists at t'', then it is not true at t''
- $\underline{A}^{\dagger} \smile B^{*}$, if expressed at time t, is true at t iff there are n *supposita* of A at t (with $n \geq 1$), and for some i, for every t', if $\underline{a}_i^{\dagger} \mapsto B$ exists at t', then it is true at t'
- $\underline{A}^{\dagger} \smile B^{\dagger}$, if expressed at time t, is true at t iff there are n *supposita* of A at t (with $n \geq 1$), and for some i, for some t', if $\underline{a}_i^{\dagger} \mapsto B$ exists at t', then it is true at t'
- $A^{\dagger} \smile B^{\$}$, if expressed at time t, is true at t iff (i) there are no *supposita* of A at t, or (ii) there are n *supposita* of A at t (with $n \geq 1$), and for some i, for every t', if $\underline{a}_i^{\dagger} \mapsto B$ exists at t', then it is not true at t'
- $\underline{A}^{\dagger} \smile B^{\ddagger 2}$, if expressed at time t, is true at t iff there are n *supposita* of A at t (with $n \geq 1$), and for some i, for some t', t'', if $\underline{a}_i^{\dagger} \mapsto B$ exists at t', then it is true at t', and if $\underline{a}_i^{\dagger} \mapsto B$ exists at t'', then it is not true at t''
- $A^{\dagger} \mapsto B^{*\sim}$, if expressed at time t, is true at t iff $A^{\dagger} \mapsto B^{\$}$ is true at t
- $A^{\dagger} \mapsto B^{\dagger\sim}$, if expressed at time t, is true at t iff (i) there are no *supposita* of A at t, or (ii) there are n *supposita* of A at t (with $n \geq 1$), and for every i, for some t', if $a_i^{\dagger} \mid B$ exists at t', then it is true at t'
- $\underline{A}^{\dagger} \mapsto B^{\$\sim}$, if expressed at time t, is true at t iff $\underline{A}^{\dagger} \mapsto B^{*}$ is true at t
- $\underline{A}^{\dagger} \mapsto B^{\ddagger 2\sim}$, if expressed at time t, is true at t iff $\underline{A}^{\dagger} \mapsto B^{\ddagger 2}$ is true at t
- $A^{\dagger} \smile B^{*\sim}$, if expressed at time t, is true at t iff $A^{\dagger} \smile B^{\$}$ is true at t
- $A^{\dagger} \smile B^{\dagger\sim}$, if expressed at time t, is true at t iff (i) there are no *supposita* of A at t, or (ii) there are n *supposita* of A at t (with $n \geq 1$), and for some i, for some t', if $a_i^{\dagger} \mid B$ exists at t', then it is true at t'
- $A^{\dagger} \smile B^{\$\sim}$, if expressed at time t, is true at t iff $\underline{A}^{\dagger} \smile B^{*}$ is true at t
- $A^{\dagger} \smile B^{\ddagger 2\sim}$, if expressed at time t, is true at t iff $\underline{A}^{\dagger} \smile B^{\ddagger 2}$ is true at t

One might worry that implications of the form 'if $\underline{a}_i^{\dagger} \mapsto B$ ($a_i^{\dagger} \mid B$) exists at t', then it is (not) true at t" are vacuously true in case $\underline{a}_i^{\dagger} \mapsto B$ ($a_i^{\dagger} \mid B$) doesn't exist at t'. This would have some rather unwanted consequences. Take the truth conditions of $\underline{A}^{\dagger} \mapsto B^{*}$ and $A^{\dagger} \mapsto B^{\$}$, for example. Assume that some substitution instances of these schemes are expressed at time t yet there is no t' where a proposition of the form $\underline{a}_i^{\dagger} \mapsto B$ exists. All instances are true at t in case A has existential import at t. That is, we would commit Fabri to the view that $\underline{A}^{\dagger} \mapsto B^{*}$ and $A^{\dagger} \mapsto B^{\$}$ can be true together, which is not only highly counterintuitive but also goes against some of the most foundational principles from Fabri's logical geometry (more on this in Sect. 6.2). Fortunately, this worry can be dispelled. The implications 'if $\underline{a}_i^{\dagger} \mapsto B$ ($a_i^{\dagger} \mid B$) exists at t', then it is (not) true at t'' are not object-linguistic material implications. They are metalinguistic implications, and thus the *ex falso quodlibet* paradox does not apply here.[15]

By Fabri's truth conditions, divided modals involving 'not possible' ($B^{\sim\dagger}$ in symbols) are always equivalent to their corresponding propositions involving 'impossible' (i.e., $B^{\S}$). His truth conditions also entail the duality of necessity and possibility, which we have seen also holds in Fabri's logic of composite modals. Here is the proof for universal-affirmative divided modals. The proofs for divided modals of a different quantity are entirely analogous.

$$\text{(f)}\quad (\underline{A}^{\dagger} \mapsto B^{\dagger}) \leftrightarrow (\underline{A}^{\dagger} \mapsto B^{\sim*\sim})$$

Proof Consider any time t and assume that $\underline{A}^{\dagger} \mapsto B^{\dagger}$, $\underline{A}^{\dagger} \mapsto B^{\sim*\sim}$ exist at t. By the truth conditions of $\underline{A}^{\dagger} \mapsto B^{\dagger}$, this proposition is true at t iff A is non-empty at t, and for every i, for some t', if $a_i^{\dagger} \mapsto B$ exists at t', then it is true at t'; that is, A is non-empty at t, and for every i, for some t', if $a_i^{\dagger} \mid B$ exists at t', then it is not true at t'; that is, iff $\underline{A}^{\dagger} \mapsto B^{\sim*\sim}$ is true at t, by the truth conditions of $\underline{A}^{\dagger} \mapsto B^{\sim*\sim}$. ∎

$$\text{(g)}\quad (\underline{A}^{\dagger} \mapsto B^{*}) \leftrightarrow (\underline{A}^{\dagger} \mapsto B^{\sim\dagger\sim})$$

Proof Consider any time t and assume that $\underline{A}^{\dagger} \mapsto B^{*}$, $\underline{A}^{\dagger} \mapsto B^{\sim\dagger\sim}$ exist at t. By the truth conditions of $\underline{A}^{\dagger} \mapsto B^{*}$, this proposition is true at t iff A is non-empty at t, and for every i, for every t', if $a_i^{\dagger} \mapsto B$ exists at t', then it is true at t'; that is, iff A is non-empty at t, and for every i, for every t', if $a_i^{\dagger} \mid B$ exists at t', then it is not true at t'; that is, iff $\underline{A}^{\dagger} \mapsto B^{\sim\dagger\sim}$ is true at t, by the truth conditions of $\underline{A}^{\dagger} \mapsto B^{\sim\dagger\sim}$. ∎

Furthermore, Fabri's truth conditions also validate a number of implications where the antecedent is a divided modal and the consequent is the divided modal's corresponding assertoric proposition, or *vice versa*, although to our knowledge Fabri nowhere discusses such implications. For instance, Fabri is committed to accept the validity of schemes (h)-(m).

---

[15]We thank an anonymous reviewer for pointing us to this possible misinterpretation. It also bears emphasizing that this kind of situation is by no means restricted to token-based semantics, but also arises in contemporary systems of formal logic. For example, in public announcement logic, the semantic clause for the public announcement operator reads as follows: $\mathcal{M}, s \vDash [\varphi]\psi$ iff $\mathcal{M}, s \vDash \varphi$ implies $\mathcal{M}|\varphi, s \vDash \psi$ (van der Hoek et al. 2008: 74). The word 'implies' in this metalinguistic statement does not express a purely truth-functional, material conditional either. After all, if its antecedent fails, i.e., $\mathcal{M}, s \nvDash \varphi$, then the state s does not belong to the model $\mathcal{M}|\varphi$ in the first place, thus rendering the consequent ($\mathcal{M}|\varphi, s \vDash \psi$) meaningless.

(h) $(\underline{a} \mapsto B^*) \rightarrow (\underline{a} \mapsto B)$
(i) $(\underline{a} \mapsto B) \rightarrow (\underline{a} \mapsto B^\dagger)$
(j) $(a \mapsto B^\S) \rightarrow (a \mid B)$
(k) $(a \mid B) \rightarrow (a \mapsto B^{\dagger\sim})$
(l) $(\underline{A} \smile B) \rightarrow (\underline{A}^\dagger \smile B^\dagger)$
(m) $(A^\dagger \mapsto B^\S) \rightarrow (A \mid B)$

The list can be extended if we bear in mind that the supposition of the subject term in a divided modal can be 'restricted' (*restricta*), as Fabri calls it. The term 'restriction' here refers to the blocking of ampliation, in the sense that an ampliated term whose supposition is restricted is taken to stand for its actual *supposita* only, and no longer for both its actual and non-actual-but-possible *supposita*. Fabri refers to restriction a number of times in the questions on the *Analytica priora* (see, e.g., the passages quoted in fn. 12 on p. 68), but he does not treat the topic in any detail. We may assume that, like most of his contemporaries, he believed that the standard way to restrict the supposition of an ampliated term A is to turn A into the antecedent of a relative clause 'that exists' (*quod est*).[16] For instance, in 'For every A that exists it is necessary to be B' (*Omne A quod est necesse est esse B*), A is restricted to supposit only for its actual *supposita*. If we assume that the restriction mechanism is available, then schemes (n)-(s) are also valid:

(n) $(\underline{A} \mapsto B) \rightarrow (\underline{A} \mapsto B^\dagger)$
(o) $(\underline{A} \mapsto B^*) \rightarrow (\underline{A} \mapsto B)$
(p) $(A \mid B) \rightarrow (A \mapsto B^{\dagger\sim})$
(q) $(A \twoheadrightarrow B) \rightarrow (A \smile B^{\dagger\sim})$
(r) $(\underline{A} \smile B^*) \rightarrow (\underline{A} \smile B)$
(s) $(A \twoheadrightarrow B^\dagger) \rightarrow (A \twoheadrightarrow B)$[17]

As the reader will have noticed, schemes (h)-(s) are all variants on schemes that are valid in present-day quantified modal logic T. Add to this that Fabri does not treat iterated modalities, and the conclusion should be that if we were to develop a Kripke-style semantics for Fabri's logic of divided modals of the kind Johnston has constructed for Buridan and Thom for Kilwardby, then reflexive models will presumably be sufficiently strong to validate whatever is valid in Fabri's original system.[18]

[16] See Ashworth (1974: 92); Broadie (1985: 81–82) on late-medieval theories of restriction.

[17] Note that, perhaps counterintuitively, $(A \twoheadrightarrow B) \rightarrow (A^\dagger \smile B^{\dagger\sim})$ and its contrapositive $(\underline{A}^\dagger \mapsto B^*) \rightarrow (\underline{A} \mapsto B)$ are invalid. To see this, consider the time t where (1) $A \twoheadrightarrow B$, $A^\dagger \smile B^{\dagger\sim}$, $\underline{A}^\dagger \mapsto B^*$, and $\underline{A} \mapsto B$ all exist, (2) A has a number of non-actual-possible *supposita*, all of which are among the *supposita* of B at every t', and (3) A does not have any actual *supposita*. Since A lacks actual *supposita*, $A \twoheadrightarrow B$ is true at t, and since all of A's possible *supposita* are among the *supposita* of B at every t', $\underline{A}^\dagger \mapsto B^*$ is also true at t. Thus, $A^\dagger \smile B^{\dagger\sim}$, which we shall see is contradictory to $\underline{A}^\dagger \mapsto B^*$, is false at t, meaning t constitutes a counterexample to the validity of $(A \twoheadrightarrow B) \rightarrow (A^\dagger \smile B^{\dagger\sim})$. Likewise, since $A \twoheadrightarrow B$ is true at t, its contradictory $\underline{A} \mapsto B$ is false at t, and so t also constitutes a counterexample to the validity of $(\underline{A}^\dagger \mapsto B^*) \rightarrow (\underline{A} \mapsto B)$. In short, both implications are invalid since $A \twoheadrightarrow B$ and $\underline{A}^\dagger \mapsto B^*$ are consistent with each other. For more on Fabri's account of the opposition relations between divided modals, see Sect. 6.2.

[18] See Hodges and Johnston (2017); Johnston (2015a); Thom (2019).

6.2 The Logical Geometry of Divided Modals

With Fabri's truth conditions of divided modals clarified, we can turn to his account of the opposition relations holding between such propositions. Fabri discusses this topic at some length in the *quaestiones* on *De int.* 12-13.[19] The relations that he considers are the standard four: contradiction, contrariety, subcontrariety, and subalternation. These relations had been common currency among logicians ever since late antiquity, and Fabri's interpretation of them is in no way remarkable. Two propositions φ and ψ.

- are contradictories ($\leftrightarrow$) iff φ and ψ cannot be true together, and φ and ψ cannot be false together
- are contraries (—) iff φ and ψ cannot be true together, and φ and ψ can be false together
- are subcontraries (- -) iff φ and ψ can be true together, and φ and ψ cannot be false together
- stand in a subalternation relation ($\rightarrow$) iff φ entails ψ, and ψ does not entail φ[20]

These equivalences are coarse-grained descriptions. Fabri endorses them, but they do not do full justice to the token-based nature of his logic. The right-hand side of each of these equivalences should, in fact, be specified further in accordance with the following three keys, which describe the notions of entailment and possible joint truth and falsity in token-based settings.

(K_1) φ and ψ can be true (false) together iff for some time t, φ and ψ are both true (false) at t if φ and ψ co-exist at t.
(K_2) φ and ψ cannot be true (false) together iff for every time t, either φ or ψ is false (true) at t if φ and ψ co-exist at t.
(K_3) φ entails ψ iff ψ is true at t if ψ exists at t, for every time t such that φ is true at t if φ exists at t.

[19] See **S**, ll. 36–81.

[20] Fabri treats these four opposition relations in the quaestiones on *De int.* 6-7, at **S**, ff. 204ra-209va. The text does not contain clear-cut characterizations of the different relations – all *quaestiones* relate to highly specific issues –, but Fabri points out on several occasions that he presupposes, and thus accepts the truth of, the '*leges propositionum oppositarum*' that are found in Peter of Spain's *Summulae logicales*; see e.g., **S**, f. 206va: "... et hic presupponitur quod triplex est materia propositionum, sicut dicit Petrus Hispanus de legibus propositionum oppositarum. Item hic presupponitur causa que solet assignari quare contrarie non possunt simul esse uere et sic de aliis, que dicta sunt in Petro Hispano." The bullet points in the running text paraphrase Peter's discussion at *Sum. log*. I.14 (De Rijk 1972: 7 [ll. 11–31]).

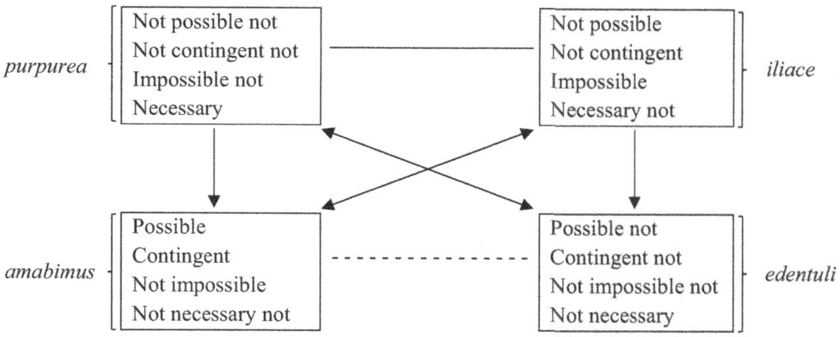

Fig. 6.1 The 'bare' modal square of opposition

Thus, for instance, we obtain a fine-grained description of contradiction if we restate the coarse-grained one by applying (K_2) twice (once to each conjunct in the coarse-grained description):

φ and ψ are contradictories	iff	φ and ψ cannot be true together, and φ and ψ cannot be false together
	iff	for every time t, if φ and ψ co-exist at t, then either φ is true at t and ψ is false at t, or ψ is true at t and φ is false at t

To be clear, Fabri is content with the coarse-grained descriptions and does not take this extra step. None of the keys (K_1)–(K_3) occurs in his commentaries. But since they do little more than finetuning some core notions in accordance with one of the most basic assumptions underpinning his logic – namely, that propositions are sentence tokens – it is safe to assume that he would have endorsed them.

While exploring which pairs of divided modals involve which opposition relation, Fabri makes extensive use of four terms: '*purpurea*', '*iliace*', '*amabimus*', and '*edentuli*'. Before we proceed, it is important that we get clear on their meaning. These four terms are technical terms, which are commonly found in discussions of modal opposition from the thirteenth century onwards. In most texts they are used to describe the so-called modal square of opposition, a figure consisting of four vertices where each vertex is an equivalence class, containing either four equivalent modal propositions, or, more frequently, four equivalent modalities (the so-called 'bare' modal square, which is drawn in Fig. 6.1).[21] In those discussions, each of the four terms indicates one vertex, and provides information about the nature of the modalities in that vertex: '*purpurea*' indicates the upper-left vertex, '*iliace*' the upper-right

[21] See, e.g., Grey (2017: 211–212) and Uckelman (2008: 395–396), who discuss the use of the four terms in resp. the *Port-Royal Logic* and some thirteenth-century texts. Also, the modal square is based on Aristotle's *De int.* 12-13, although it is not of Aristotle's own invention; see Geudens and Demey (2021). Note that 'contingent' is used in its one-sided sense in the modal square, as equivalent to 'possible'.

vertex, '*amabimus*' the lower-left vertex, and '*edentuli*' the lower-right vertex. Each term consists of four syllables, and each syllable concerns one modality: the first syllable in each term concerns possibility, the second one contingency, the third one impossibility, and the fourth one necessity. Each syllable, moreover, contains exactly one vowel. This vowel specifies whether there are negations surrounding the modality that the syllable in which the vowel occurs, is concerned with:

- *a*: the modality is neither preceded nor followed by a negation
- *e*: the modality is followed, but not preceded, by a negation
- *i*: the modality is preceded, but not followed, by a negation
- *u*: the modality is both preceded and followed by a negation

For instance, '*purpurea*' indicates that the upper-left corner of the square contains the following list of equivalent modalities: 'not possible not' (*u*, applied to 'possible'), 'not contingent not' (*u*, applied to 'contingent'), 'impossible not' (*e*, applied to 'impossible'), 'necessary' (*a*, applied to 'necessary').

In later fifteenth-century discussions of modal opposition, however, these terms are occasionally used outside the context of the modal square, and in a much broader sense. On this second interpretation, they are still associated with equivalence classes of modalities, as is also the case on the received view, but as opposed to the received view, the terms are not used to denote bare modalities. Rather, each term now picks out all divided modal propositions that involve one of the modalities in this equivalence class. For instance, '*purpurea*' is still associated with the equivalence class consisting of 'necessary', 'impossible not', 'not possible not', and 'not contingent not', but the term now picks out *all* divided modals involving one of these modalities, including, for example 'For every *A* it is necessary to be *B*', 'For some *A* it is impossible not to be *B*', and so on. This wider usage is attested in Coronel, Crockaert, and Dullaert, among others, and Fabri, too, uses the terms in this way.[22] He addresses their meaning in the passage below, which immediately follows on the passage in quote [*l*]:

> [*p*] It is also clear that all *purpurea* propositions are universal with respect to the modality and affirmative (*uniuersales ex parte modi et affirmatiue*); that all *iliace* propositions are universal and negative; that all *amabimus* propositions are particular and affirmative; and that all *edentuli* propositions are particular and negative, and this with respect to their modality (*et hoc ex parte modi*).[23]

Fabri's characterization in quote [*p*] is terse, but the passage does not pose too many difficulties if we bear in mind what Fabri has to say about the quality and modal quantity of divided modals: if characterized up to logical equivalence, then *purpurea* propositions are all and only divided modals of necessity; *iliace* propositions are all and only divided modals of impossibility; *amabimus* propositions are all and only divided modals of possibility; and *edentuli* propositions are all and only divided

[22] See, e.g., Coronel, *Prima pars rosarii* (Coronel 1517: sig. g ii^{va} - iii^{va}); Crockaert, *in Sum. log.* (Crockaert 1512: sig. d iiii$^{va\text{-}b}$); Dullaert, *in De int.* (Dullaert 1515: ff. 130ra-131vb).
[23] S, ll. 32–35.

modals involving 'possible not' (the O-corner of the modal square not being lexicalized). Thus, the following is an adequate extensional characterization, again up to logical equivalence, of the meaning of the four terms in Fabri's logic for divided modals:

- the *supposita* of '*purpurea*' at time *t* are the propositions $\underline{A}^{\dagger} \mapsto B^*$, $\underline{A}^{\dagger} \smallsmile B^*$, and $\underline{a} \mapsto B^*$ existing at *t*
- the *supposita* of '*iliace*' at time *t* are the propositions $A^{\dagger} \mapsto B^{\S}$, $A^{\dagger} \smallsmile B^{\S}$, and $a \mapsto B^{\S}$ existing at *t*
- the *supposita* of '*amabimus*' at time *t* are the propositions $\underline{A}^{\dagger} \mapsto B^{\dagger}$, $\underline{A}^{\dagger} \smallsmile B^{\dagger}$, and $\underline{a} \mapsto B^{\dagger}$ existing at *t*
- the *supposita* of '*edentuli*' at time *t* are the propositions $A^{\dagger} \mapsto B^{\dagger\sim}$, $A^{\dagger} \smallsmile B^{\dagger\sim}$, and $a \mapsto B^{\dagger\sim}$ existing at *t*

With these terminological preliminaries out of the way, we can move on and reconstruct Fabri's thoughts on the opposition relations between divided modals. In what follows we shall leave singular divided modals out of consideration, and focus on universal and particular divided modals instead, as they are the main concern of Fabri himself and their case is the more interesting one. Fabri sets out his account in four rules, or theorems (*regulae*), which we shall discuss in turn. The first two rules address the relations of resp. contrariety and subcontrariety:

> [*q*] The first rule is that [i] *purpurea* and *iliace* propositions are contrary to each other if they are both universal with respect to the subject or at least one of them is. This is proven, for such propositions have the same subject and predicate, and one is affirmative and the other is negative, and both are universal – with respect to the modality, that is – and they satisfy the law of contraries (*participant legem contrariarum*) ... But [ii] if both propositions are particular with respect to their subject, then they are unconnected (*disparate*); that is, they are not opposed to each other in any way (*nullo modo opposite*) ... The second rule is that [iii] *amabimus* and *edentuli* propositions with the same subject and predicate are subcontrary to each other, if they are both particular with respect to the subject or at least one of them is. But [iv] if they are both universal with respect to the subject, then they are unconnected ... From these rules it also seems to follow (*sequi uidetur*) that [v] the *purpurea* proposition of necessity that is universal with respect to the subject and the *edentuli* proposition of possibility that is universal with respect to the subject are contrary to each other ... From the second rule it also follows that [vi] the *purpurea* proposition of necessity and the *edentuli* proposition of possibility are subcontrary to each other if they are both particular with respect to the subject ...[24]

If we read this passage in the light of the above characterization of the meaning of '*purpurea*', '*iliace*', '*amabimus*' and '*edentuli*', then in rules one ([i]-[ii]) and two ([iii]-[iv]) Fabri is claiming that

- [i] $\underline{A}^{\dagger} \mapsto B^*$ and $A^{\dagger} \mapsto B^{\S}$ are contrary to each other
- [i] $\underline{A}^{\dagger} \mapsto B^*$ and $A^{\dagger} \smallsmile B^{\S}$ are contrary to each other

[24]**S**, ll. 37–57. Note that the additions 'of necessity' (*de necessario*) and 'of possibility' (*de possibili*) in the descriptions of the *purpurea* and *edentuli* propositions at [v] and [vi] are strictly speaking superfluous. *Purpurea* propositions by definition involve 'necessary', and *edentuli* propositions by definition involve 'possible not'.

- [i] $\underline{A}^{\dagger} \smallsmile B^{*}$ and $A^{\dagger} \mapsto B^{\S}$ are contrary to each other
- [ii] $\underline{A}^{\dagger} \smallsmile B^{*}$ and $A^{\dagger} \smallsmile B^{\S}$ are unconnected
- [iii] $\underline{A}^{\dagger} \smallsmile B^{\dagger}$ and $A^{\dagger} \smallsmile B^{\dagger\sim}$ are subcontrary to each other
- [iii] $\underline{A}^{\dagger} \smallsmile B^{\dagger}$ and $A^{\dagger} \mapsto B^{\dagger\sim}$ are subcontrary to each other
- [iii] $\underline{A}^{\dagger} \mapsto B^{\dagger}$ and $A^{\dagger} \smallsmile B^{\dagger\sim}$ are subcontrary to each other
- [iv] $\underline{A}^{\dagger} \mapsto B^{\dagger}$ and $A^{\dagger} \mapsto B^{\dagger\sim}$ are unconnected

At [v] and [vi], Fabri makes two further claims, one about contrariety and one about subcontrariety:

- [v] $\underline{A}^{\dagger} \mapsto B^{*}$ and $A^{\dagger} \mapsto B^{\dagger\sim}$ are contrary to each other
- [vi] $\underline{A}^{\dagger} \smallsmile B^{*}$ and $A^{\dagger} \smallsmile B^{\dagger\sim}$ are subcontrary to each other

All claims are true if we assume that the truth conditions of Fabri's divided modals are as reconstructed in Sect. 6.1., as the reader can check. Fabri's third and fourth rule focus primarily on contradiction and subalternation, but also yield one more contrariety, one more subcontrariety, and two more unconnected pairs:

> [*r*] The third rule is that [i] *purpurea* and *edentuli* propositions with the same subject and predicate are contradictory to each other if one of them is particular with respect to the subject and the other is universal. Likewise, [ii] *iliace* and *amabimus* propositions with the same subject and predicate are contradictory to each other if one is universal with respect to the subject and the other is particular with respect to the subject. This is proven, for such propositions have the same subject and predicate, one is affirmative and the other is negative, they are of different quantity – both with respect to the modality and with respect to the subject – and they satisfy the law of contradictories (*participant legem contradictoriorum*) ... But [iii] if such propositions are both universal with respect to the subject, then they are contrary to each other; and [iv] if they are both particular with respect to the subject, then they are subcontraries ... The fourth rule is that [v] *amabimus* propositions are subaltern to *purpurea* propositions with the same subject and predicate, if there is more universality in one than in the other (*si sit plus de uniuersalitate in una quam in altera*). From this it follows that the *amabimus* proposition that is universal with respect to the subject is subaltern to the *purpurea* proposition that is universal with respect to the subject, if they have the same subject and predicate. Likewise, the particular *amabimus* proposition is subaltern to both the *purpurea* proposition that is universal with respect to the subject and the *purpurea* proposition that is particular with respect to the subject. [vi] The particular *purpurea* proposition and the universal *amabimus* proposition are unconnected, and do not satisfy any law of opposition (*non habentes aliquam legem oppositionis*). [vii] Likewise for *iliace* and *edentuli* propositions (*similiter est in iliace et edentuli*).[25]

In the third and fourth rule, Fabri is making the following claims, all of which are true if we assume the truth conditions from Sect. 6.1.

- [i] $\underline{A}^{\dagger} \mapsto B^{*}$ and $A^{\dagger} \smallsmile B^{\dagger\sim}$ are contradictory to each other
- [i] $\underline{A}^{\dagger} \smallsmile B^{*}$ and $A^{\dagger} \mapsto B^{\dagger\sim}$ are contradictory to each other
- [ii] $\underline{A}^{\dagger} \smallsmile B^{\dagger}$ and $A^{\dagger} \mapsto B^{\S}$ are contradictory to each other
- [ii] $\underline{A}^{\dagger} \mapsto B^{\dagger}$ and $A^{\dagger} \smallsmile B^{\S}$ are contradictory to each other
- [iii] $\underline{A}^{\dagger} \mapsto B^{\dagger}$ and $A^{\dagger} \mapsto B^{\S}$ are contrary to each other

[25] S, ll. 60–77.

- [iv] $\underline{A}^{\dagger} \smile B^{\dagger}$ and $A^{\dagger} \smile B^{\S}$ are subcontrary to each other
- [v] $\underline{A}^{\dagger} \mapsto B^{\dagger}$ is subaltern to $\underline{A}^{\dagger} \mapsto B^{*}$
- [v] $\underline{A}^{\dagger} \smile B^{\dagger}$ is subaltern to $\underline{A}^{\dagger} \mapsto B^{*}$
- [v] $\underline{A}^{\dagger} \smile B^{\dagger}$ is subaltern to $\underline{A}^{\dagger} \smile B^{*}$
- [vi] $\underline{A}^{\dagger} \smile B^{*}$ and $\underline{A}^{\dagger} \mapsto B^{\dagger}$ are unconnected
- [vii] $A^{\dagger} \mapsto B^{\dagger\sim}$ is subaltern to $A^{\dagger} \mapsto B^{\S}$
- [vii] $A^{\dagger} \smile B^{\dagger\sim}$ is subaltern to $A^{\dagger} \mapsto B^{\S}$
- [vii] $A^{\dagger} \smile B^{\dagger\sim}$ is subaltern to $A^{\dagger} \smile B^{\S}$
- [vii] $A^{\dagger} \mapsto B^{\dagger\sim}$ and $A^{\dagger} \smile B^{\S}$ are unconnected

Rules one to four exhaust Fabri's account of the opposition relations between divided modals of either particular or universal quantity. That means that his account is not entirely complete. The relations that Fabri describes entail four further subalternation relations, which however are not mentioned in Fabri's text. The issue concerns the following implication, which is a theorem in Fabri's logic of assertoric propositions and, by extension, also in his logic of divided modals:

(S-Theorem) For any divided modals of particular or universal quantity φ_1, φ_2, φ_3 and φ_4, if φ_1 is contrary to φ_2, φ_1 is contradictory to φ_3, and φ_2 is contradictory to φ_4, then φ_4 is subaltern to φ_1 and φ_3 is subaltern to φ_2.

Proof Assume that φ_1 is contrary to φ_2, that φ_1 is contradictory to φ_3, and that φ_2 is contradictory to φ_4. We now show:

1. φ_4 is subaltern to φ_1: Assume that φ_1 is true. Then φ_2 is false, since contraries cannot be true together; and φ_4 is true, since contradictories cannot be false together. So φ_1 entails φ_4. But φ_4 does not entail φ_1. After all, the contraries φ_1 and φ_2 can be false together, i.e., it is possible that φ_1 is false and φ_2 is false. Since φ_2 and φ_4 are contradictories and thus cannot be false together, it follows that it is possible that φ_1 is false and φ_4 is true, and thus, that φ_4 does not entail φ_1.
2. φ_3 is subaltern to φ_2: The proof is entirely analogous to (1). ∎

The S-theorem establishes that for each of the proposition clusters *purpurea*, *iliace*, *amabimus* and *edentuli*, there are subalternation relations between the universal and the particular propositions in these clusters:

- $\underline{A}^{\dagger} \mapsto B^{*}$ is contrary to $A^{\dagger} \mapsto B^{\dagger\sim}$ ([v], quote [*q*]), $\underline{A}^{\dagger} \mapsto B^{*}$ is contradictory to $A^{\dagger} \smile B^{\dagger\sim}$, and $A^{\dagger} \mapsto B^{\dagger\sim}$ is contradictory to $\underline{A}^{\dagger} \smile B^{*}$ ([i], quote [*r*]). Thus, by the S-theorem, $\underline{A}^{\dagger} \smile B^{*}$ is subaltern to $\underline{A}^{\dagger} \mapsto B^{*}$, and $A^{\dagger} \smile B^{\dagger\sim}$ is subaltern to $A^{\dagger} \mapsto B^{\dagger\sim}$.
- $\underline{A}^{\dagger} \mapsto B^{\dagger}$ is contrary to $A^{\dagger} \mapsto B^{\S}$ ([iii], quote [*r*]), $\underline{A}^{\dagger} \mapsto B^{\dagger}$ is contradictory to $A^{\dagger} \smile B^{\S}$, and $A^{\dagger} \mapsto B^{\S}$ is contradictory to $\underline{A}^{\dagger} \smile B^{\dagger}$ ([ii], quote [*r*]). Thus, by the S-theorem, $\underline{A}^{\dagger} \smile B^{\dagger}$ is subaltern to $\underline{A}^{\dagger} \mapsto B^{\dagger}$, and $A^{\dagger} \smile B^{\S}$ is subaltern to $A^{\dagger} \mapsto B^{\S}$.

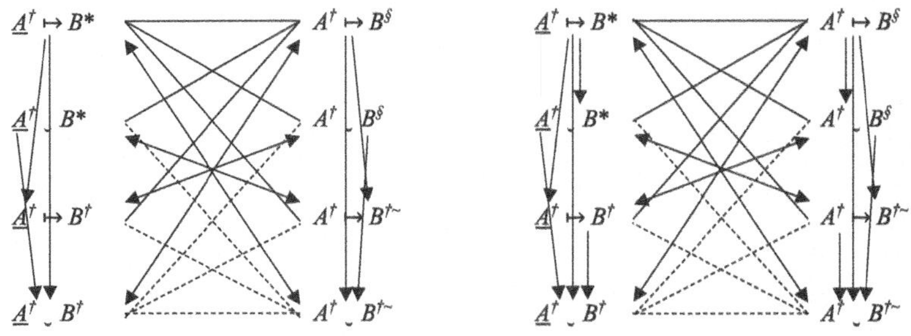

Fig. 6.2 Fabri's account of modal opposition visualized

None of these relations is listed under rule four in quote [*r*], or anywhere else in Fabri's discussion of modal opposition. That Fabri forgot about precisely these four relations is probably a consequence of his very specific approach to the topic. Fabri's focus is not on eight proposition types, as one would expect, but rather on four *pairs of* proposition types, viz. the *purpurea* pair, the *amabimus* pair, the *iliace* pair, and the *edentuli* pair. Given this 'pairwise' approach, it is understandable for him to discuss all relations between proposition types belonging to *different* pairs, yet to forget about the relations between proposition types *inside* one and the same pair. The relations that Fabri describes in the four rules in quotes [*q*] and [*r*] can be visualized as an octagon of opposition like the one that is displayed on the left in Fig. 6.2. If we add the four subalternations that Fabri ommitted but follow from his account by the S-theorem, then we get the completed octagon on the right in Fig. 6.2.[26] Note that there are good reasons to believe that Fabri's four rules are indeed intended as a description of such an octagon. **S** contains a modal octagon on f. 223^{v} (see Fig. 6.3), which is inserted right between the *quaestiones* on *De interpretatione* and *Analytica priora*; and in the course of his discussion of the four rules Fabri refers to a 'figure on the opposition of modal propositions' (*figura de oppositionibus modalium*), which is almost certainly a reference to the octagon on f. 223^{v}.[27]

By way of conclusion, the main elements of the discussion in this chapter are summarized in statements $\mathbf{A}_{9\text{–}14}$ below. These should be read in connection with statements $\mathbf{A}_{1\text{–}3}$ on the foundations of Fabri's modal logic (Chap. 2), and $\mathbf{A}_{4\text{–}8}$ on Fabri's logic of composite modals (Chaps. 3 and 4).

- $\mathbf{A}_9$: A proposition is a divided modal if it is an instance of one of the templates in T_{DM}.
- $\mathbf{A}_{10}$: Affirmative divided modals have existential import, negative divided modals do not.

[26]The arrows in Fig. 6.2 are used in the same sense as in Fig. 6.1: double-headed arrows indicate contradiction, single-headed arrows indicate subalternation, dotted lines subcontrariety, and full lines contrariety.

[27]**S**, l. 59.

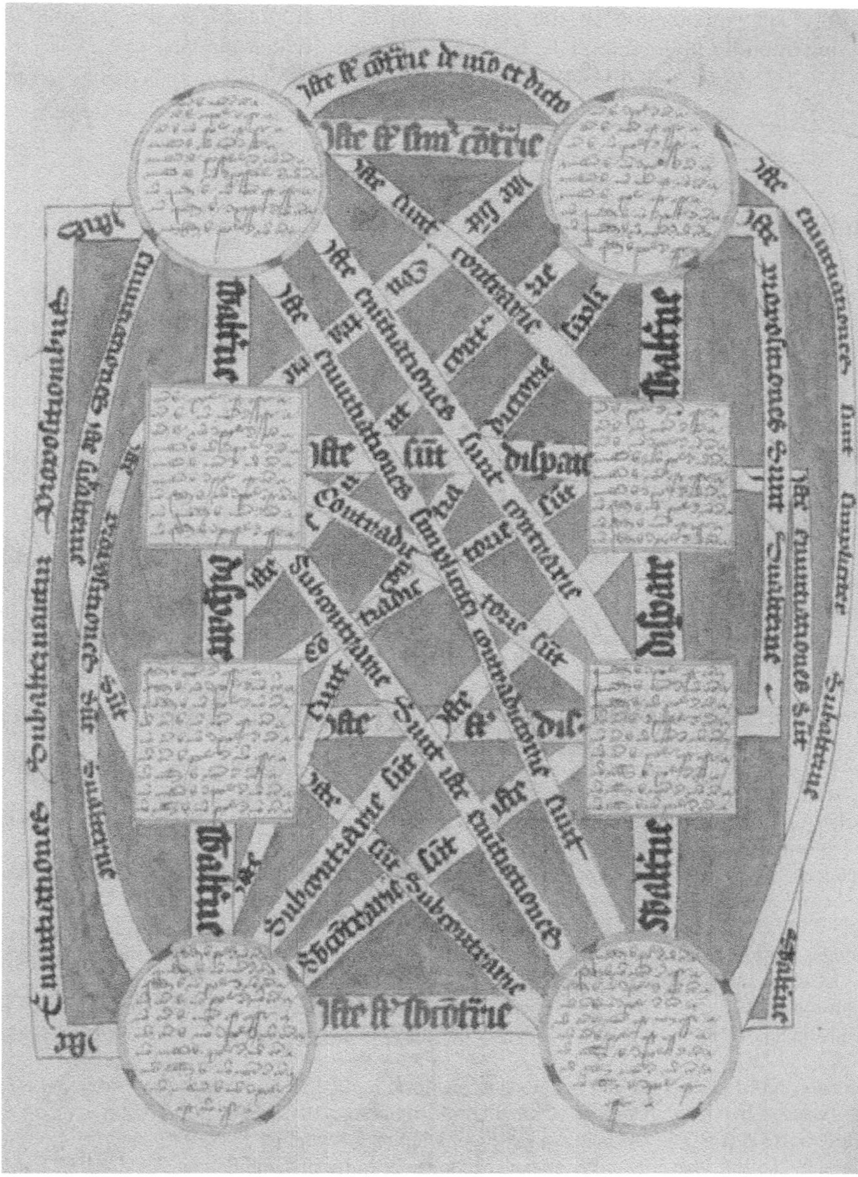

Fig. 6.3 Fabri's modal octagon as it occurs on **S**, f. 223^{v}

- $\mathbf{A}_{11}$: The truth conditions of singular divided modals are analyzed by *positio inesse* involving reduction to singular assertoric propositions.
- $\mathbf{A}_{12}$: The truth conditions of non-singular divided modals are analyzed by *positio inesse* involving reduction to sets of singular (deictic) assertoric propositions.

- A_{13}: The supposition of the logical subject is ampliated to the possible in non-singular divided modals, but not in singular divided modals.
- A_{14}: The opposition relations between non-singular divided modals can be geometrized as an octagon.

References

Ashworth EJ (1973) Existential assumptions in late medieval logic. Am Philos Q 10:141–147

Ashworth EJ (1974) Language and logic in the post-medieval period. Reidel, Dordrecht

Ashworth EJ (1977) Chimeras and imaginary objects: a study in the post-medieval theory of signification. Vivarium 15:57–77

Broadie A (1985) The circle of John Mair. Logic and logicians in pre-reformation Scotland. Clarendon Press, Oxford

Church A (1965) The history of the question of existential import of categorical propositions. In: Bar-Hillel Y (ed) Logic, methodology and philosophy of science. Proceedings of the 1964 international congress. North-Holland, Amsterdam, pp 417–424

Ciola G (2020) Hic sunt chimaerae? On absolutely impossible significates and referents in mid-14th-century nominalist logic. Recherches de Théologie et Philosophie médiévales 87: 441–467

Coombs JS (1990) John Mair and Domingo de Soto on the reduction of iterated modalities. In: Angelelli I, d'Ors A (eds) Estudios de historia de la logica. Ediciones Eunate, Pamplona, pp 161–181

Coronel A (1517) Prima pars rosarii. Venalia apud Oliuerium Senant, [Paris]

Crockaert P (1512) Summularum artis dialectice vtilis admodum interpretatio super textum magistri Petri Hispani. Per Johannem Cleyn, Lugduni

De Rijk LM (ed) (1972) Peter of Spain (Petrus Hispanus Portugalensis): *Tractatus*, called afterwards *Summule logicales*. First critical edition from the manuscripts. Van Gorcum, Assen

Dullaert J (1515) Questiones super duos libros Peri hermenias Aristotelis. Impresse per Stephanum Baland, [Lyon]

Geudens C, Demey L (2021) On the Aristotelian roots of the modal square of opposition. Logique et Analyse 255:303–338

Grey J (2017) The modal equivalence rules of the Port-Royal Logic. Hist Philos Log 38:210–221

Hodges W, Johnston S (2017) Medieval modalities and modern methods: Avicenna and Buridan. IfCoLog J Log Appl 4:1029–1073

Johnston S (2015a) A formal reconstruction of Buridan's modal syllogism. Hist Philos Log 36:2–17

Johnston S (2015b) Ockham and Buridan on the ampliation of modal propositions. Br J Hist Philos 23:234–255

Klima G (2001) Existence and reference in medieval logic. In: Morscher E, Hieke A (eds) New essays in free logic in honour of Karel Lambert. Springer, Dordrecht, pp 197–226

Priest G, Read S (1981) Ockham's rejection of ampliation. Mind 90:274–279

Thom P (2019) Robert Kilwardby's science of logic. A thirteenth-century intensional logic. Leiden, Brill

Uckelman SL (2008) Three 13th-century views of quantified modal logic. Adv Mod Log 7:389–406

van der Hoek W et al (2008) Dynamic epistemic logic. Springer, Dordrecht

Chapter 7
Fabri's Logic of Divided Modals in Its Historical Context

Abstract This chapter focuses on the historical context of Fabri's logic of divided modals. We are concerned with two topics: Fabri's modal octagon, and the method that he uses to determine the truth conditions of divided modals. As in Chaps. 2 and 5, we will argue that there is a clear link with the Paris *via moderna* and, indirectly, with the work of John Buridan and his associates.

Keywords John Fabri of Valenciennes · Divided modals · Modal octagon · John Mair · John Buridan · *Via moderna*

In Chap. 6, we discussed the basic tenets of Fabri's logic of divided modals. In this chapter, we will situate this logic in its historical context. As in Chap. 5, our main aim is to chart parallels with other modal logics from the period, rather than to focus on the question of originality. Fabri's truth conditions of divided modals, which we discussed in Sect. 6.1, are rather unremarkable. They are in line with the received view of the topic during the post-medieval period, among both traditionalists and modernists. By contrast, Fabri's procedure of analyzing divided modals by means of (sets of) singular assertoric propositions is more interesting, and so is his account of the logical geometry of quantified divided modals, outlined in Sect. 6.2. These are the subjects that we will discuss in this chapter. We shall see that again there is a link with the Paris *via moderna*.

Fabri's semantics essentially reduces the truth value of divided modals to a matter of the co-supposition between names of assertoric propositions and modal predicates. There are only a couple of post-medieval authors who propose a similar semantics, and all of them seem to have had ties with Mair's circle in Paris. Caubraith is a case in point. Caubraith writes that in order to assess the truth or falsity of a divided modal proposition, 'it should be reduced to one or more assertoric propositions' (*reducenda est ad unam de inesse vel ad plures*) whose membership in the *supposita* of the proposition's modality should be assessed, and he then proceeds to develop a semantics that is essentially identical to the one found in Fabri.[1]

[1] Caubraith, *Quadrupertitum* (Caubraith 1516: f. 124rb).

C. Geudens, L. Demey, *The Modal Logic of John Fabri of Valenciennes (c. 1500)*, SpringerBriefs in Philosophy, https://doi.org/10.1007/978-3-030-98802-9_7

Likewise, Pardo points out that 'any true divided modal proposition is reducible to – that is, its truth can be made clear by – one or more true or false assertoric propositions that are necessary or contingent, possible or impossible, depending on the exigencies of the mode occurring in the divided modal in question'.[2] We find analogous proposals in authors such as Celaya, Coronel, Soto, Dolz, Enzinas, and Mair.[3] This approach to divided modals goes back to fourteenth-century nominalism. It occurs fully fledged in Ockham, possibly for the first time, and later it also shows up in Albert of Saxony.[4] A fragment from his *Perutilis logica* is printed below, as quote [*s*]. Moreover, there is at least one passage, from the *Tractatus de consequentiis*, suggesting that Buridan, too, was familiar with it, and the same probably holds of Dorp.[5]

> [*s*] For the truth of a modal proposition in the divided sense ... it suffices that the mode can be verified of the proposition that consists of a pronoun demonstrating a *suppositum* of the subject of the proposition corresponding to the modal proposition's *dictum*, and that also consists of the predicate of the modal proposition but taken in its appropriate form. For instance, for the truth of this proposition 'For something white it is possible to be black' (*album possibile est esse nigrum*) it is not required that the proposition 'Something white is black' is possible, but it suffices that the proposition 'This is black' is possible, where 'This' picks out a *suppositum* of the term 'white' ...[6]

Also Fabri's account of the opposition relations between quantified divided modals shows that his modal logic is specifically modernist in inspiration. As we have seen, Fabri geometrizes these opposition relations as an octagon. We know of several other authors who were active during the *Wegestreit* period and either describe, or both describe and depict, an octagon of opposition for quantified divided modals. They were all adherents of the *via moderna*:

- John Mair describes and depicts a modal octagon in his *Quaestiones* on *De interpretatione*, which appeared in one edition, in 1528.[7]

[2]Pardo, *Medulla dyalectices* (Pardo 1505: f. 107va): "Quelibet modalis diuisa vera est reducibilis (id est, manifestabilis sua veritas) per vnam vel per plures de inesse necessarias vel contingentes, possibiles vel impossibiles, veras vel falsas, secundum exigentiam modorum positorum in tali modali diuisa."

[3]De Celaya, *in Sum. log.* (de Celaya 1525: sig. N i$^{ra\text{-}b}$); Coronel, *Prima pars Rosarii* (Coronel 1517: sig. g iii$^{ra\text{-}b}$, g v$^{ra\text{-}b}$, g vi$^{ra\text{-}b}$); Dolz, *in Sum. log.* (Dolz 1512: sig. T i^{vb}); Enzinas, *in Sum. log.* (Enzinas 1528: ff. 34rb-35rb); Major, *Introductorium perutile* (Major 1527: f. 68va); Soto, *in Sum. log.* (Soto 1529: f. 65$^{ra\text{-}va}$). See also Alt and Lagerlund (forthcoming); along with the comments in Broadie (1985: 88).

[4]See de Saxonia, *Perutilis logica* III.5 (Berger 2010: 454–458); Ockham, *Summa logicae* II.9–10 (Boehner, Gál and Brown 1974: 273–279). On Ockham, see esp. Dutilh Novaes (2004).

[5]Buridanus, *Tractatus de consequentiis* II.7 (Hubien 1976: 76–77 [ll. 214–232]); Dorp, *Compendium perutile* (Dorp 1499a: sig d 4vb).

[6]De Saxonia, *Perutilis logica* III.5 (Berger 2010: 454 [ll. 10–20]).

[7]Major, *in De int.* (Major 1528: ff. 57rb-58^{r}). All bibliographical data in this and the following bullet points are taken from Risse (1998).

- Juan Dolz describes, but does not depict, a modal octagon in his *Disceptationes* on Peter of Spain's *Summulae*, which were published only once, in 1512.[8] Dolz was a student of Gaspar Lax (1487–1560), himself a student of Mair, at the Collège de Montaigu.
- Domingo de Soto describes and depicts a modal octagon in the *Summulae*, both in the 1529 *editio princeps* and in later revisions.[9] Soto was a student of Mair at Montaigu.
- Jan Dullaert describes a modal octagon in his *Quaestiones* on *De interpretatione*, which appeared in four editions (Paris, 1509 and 1519; Lyon, 1515; Salamanca, 1517). The 1509 and 1515 editions also contain a visualization of the octagon.[10] (The 1517 and 1519 editions were not accessible to us.) Dullaert studied with Mair, and had Lax among his students.
- Robert Caubraith describes a modal octagon in his *Quadrupertitum in oppositiones, conversiones, hypotheticas et modales*, which first appeared in 1510, with a reedition in 1516.[11] Neither edition contains a visualization of the octagon. Caubraith was a student of Mair.
- The anonymous *Compilatores* describe a modal octagon in the *Compilatio ex Buridano, Dorp, Ockan et aliis nominalibus in textum Petri Hyspani*. The commentary is known to have been published three times: twice in Lyon (1495, 1500), and once in Paris (1499). Neither Lyon edition contains a visualization of the octagon. (The Paris edition was not accessible to us.) As the title shows, these authors were *moderni*.[12]
- John Dorp describes a modal octagon in his commentary on Buridan's *Summulae de dialectica*. The work appeared in at least ten editions between 1487 and 1510, the majority of which was printed in Lyon: 1487, 1490, 1493, 1495 (all in Lyon), 1499 (Venice and Lyon), 1504 (Paris), 1510 (Lyon). Those editions that were accessible to us, all contain a visualization of the octagon as well.[13]
- Peter Crockaert describes a modal octagon in the *Summularum interpretatio*, a commentary on Peter of Spain that appeared in two editions, in 1508 (Paris) and 1512 (Lyon). The 1512 edition does not contain a visualization; the 1508 edition was not accessible to us.[14] Crockaert was an adherent of Thomism, but he studied with Mair at Montaigu.

These authors constitute the immediate background against which Fabri's account of modal opposition should be understood. It deserves to be pointed out, however, that none of them adopted the pairwise approach that we find in Fabri. Most of them

---

[8] Dolz, *in Sum. log.* (Dolz 1512: sig. U vii^{va}- viii^{ra}).

[9] Soto, *Summulae* (Soto 1529: f. 54$^{va\text{-}b}$; 1554: ff. 76ra-77rb).

[10] Dullaert, *in De int.* II (Dullaert 1509: ff. 133va-136rb; 1515: ff. 121va-124rb).

[11] Caubraith, *Quadrupertitum* (Caubraith 1510: sig. S iiii$^{va\text{-}b}$; 1516: sig. S iiii$^{va\text{-}b}$).

[12] [Compilatores], *in Sum. log.* ([Compilatores] [1495]: sig. c i$^{ra\text{-}vb}$; 1500: ff. 36va-37ra).

[13] Dorp, *Perutile compendium* (Dorp 1495: sig. d i^{r}; 1499a: sig. d 2vb-3vb; 1499b: sig. d ii^{vb}- iiii^{r}; 1504: sig. c ii^{rb}-iii^{vb}).

[14] Crockaert, *in Sum. log.* (Crockaert 1512: sig. d v$^{va\text{-}b}$).

appear to have followed John Dorp, and Dorp simply takes the eight proposition types as primitives, and does not mold these types into a pairwise structure by means of the *purpurea* terminology. We should be careful with originality claims in view of the fact that so little is known about modal logic from the *Wegestreit* period, but it does seem safe to say that Fabri's approach towards the modal octagon is highly unusual. Moreover, the depictions of the modal octagon – whether handwritten or printed – that occur in the work of Fabri and several authors from the list above, testify to the increasing tendency to visualize knowledge that Ong has shown is one of the distinctive features of the philosophical discourse of the period, and that, fueled by the development of printing, culminated in the Ramist movement during the second half of the sixteenth century.[15] The late fifteenth and early sixteenth centuries were – in Northern Europe at least – characterized by a surge in interest of 'geometrizing' logical concepts, and this period also witnessed the introduction of several new kinds of logic diagrams, ranging from squares charting the relations between conjunctions, disjunctions and their negations over octagons for exponible propositions to (do)decagons for hypothetical propositions.[16] The intricate details of this phenomenon still await further study, but what matters here is that the visualizations of the modal octagon found in the sources that we are concerned with, fits within a larger development taking place in Northern European philosophy at the eve of the breakthrough of humanism.

Dorp and these later *moderni*, Fabri included, were all drawing on Buridan – Dorp directly, the later authors presumably indirectly. Buridan's account of the truth conditions of quantified divided modals, which has received ample attention in the scholarship, is also predicated on the concept of ampliation, and it is probably the ultimate source of the accounts that we find in *Wegestreit* modernists.[17] Buridan, moreover, gave a complete description of the opposition relations between quantified divided modals in the *Summulae de propositionibus* and the *Quaestiones* on *De interpretatione*, and several manuscripts of these works also contain a visualization of the modal octagon.[18] Importantly, Buridan was presumably the first author to *both* get all the relations of contradiction, contrariety, subcontrariety and subalternation between such propositions correct, *and* see that these relations can be visualized as an octagon.[19] Buridan's insight did not come out of the blue, to be sure. We know that it was anticipated by authors like Abelard, in the West, and al-Fārābī (872–950), Avicenna (c. 970–1037), and al-Kātibī (1203–1277), in the East. Thom points out that Abelard's semantics of quantified divided modals (or modal propositions *de*

[15] See Ong (1958), and, more recently, Berger (2017).

[16] See, e.g., Noreña (1975: 19–20) (on Celaya); d'Ors (1981) (on Siliceo).

[17] On Buridan's truth conditions of divided modals, see the studies listed in fn. 17 on p. 57.

[18] Buridanus, *in De int.* II, q. 9 (van der Lecq 1983: 83–91); *Summulae de dialectica* I.8.6 (van der Lecq 2005: 92–98). On Buridan's modal octagon, see esp. Demey (2019) and the literature cited there. Read (2015: 33) gives a reproduction of the modal octagon occurring in one of the manuscripts of the *Summulae de propositionibus*.

[19] Thom (2003: 171).

rebus, as Abelard calls them) 'supports the construction of two modal squares of opposition', yet he adds that Abelard himself was 'unable correctly to collect them' into a single octagon.[20] According to Chatti, something similar was going on in al-Fārābī, whose modal logic does contain 'the vertices of the modal octagon . . . but not all the relations between these vertices';[21] as well as in Avicenna: a modal octagon is 'deducible from' his modal logic, but there is no evidence suggesting that Avicenna himself was aware of this.[22] Strobino and Thom point out that the same holds for al-Kātibī, who stood under the influence of Avicenna.[23] All these authors developed a semantics that supports a modal octagon for quantified divided modals, but none of them was able to grasp the geometry behind the logic. Buridan did, and for this reason he takes a special place in the history of the modal octagon.

References

Alt G, Lagerlund H (forthcoming) The modal semantics of John Mair

Berger H (ed, trans) (2010) Albert von Sachsen: *Logik*. Felix Meiner, Hamburg

Berger S (2017) The art of philosophy. Visual thinking in Europe from the late Renaissance to the early Enlightenment. Princeton University Press, Princeton, NJ

Boehner P, Gál G, Brown S (eds) (1974) Guillelmi de Ockham Summa logicae. The Franciscan Institute, St Bonaventure

Broadie A (1985) The circle of John Mair. Logic and logicians in pre-reformation Scotland. Clarendon Press, Oxford

Caubraith R (1510) Quadrupertitum in oppositiones, conuersiones, hypotheticas et modales. In edibus Ascensianis et Ioannis Grandisiunci, Parrhisiis

Caubraith R (1516) Quadrupertitum in oppositiones, conuersiones, hypotheticas et modales. In aedibus Iodoci Badii et Emundi Fabri, Parrhisiis

Chatti S (2014) Avicenna on possibility and necessity. Hist Philos Logic 35:332–353

Chatti S (2019) Arabic logic from Al-Farabi to Averroes.A study of the early Arabic categorical, modal, and hypothetical syllogistics. Springer, Cham

[Compilatores] [1495] Compilatio ex Buridano, Dorp, Ockan et aliis nominalibus in textum Petri Hyspani edita in regali collegio Nauerre Parisius. [Nicolaus Wolff], [Lyon]

[Compilatores] (1500) Compilatio ex Buridano, Dorp, Okan et aliis nominalibus in textum Petri Hyspani edita in regali collegio Nauarre Parisius nuper a multis mendis emendata et cum additionibus Magistri Nicolai Amantis ac plurium aliorum auctorum. Impressa per Johannem Piuard, Lugduni

Coronel A (1517) Prima pars rosarii. Venalia apud Oliuerium Senant, [Paris]

---

[20] Thom (2003: 54, 56). See also Lagerlund (2000: 37–38) on this point: "There is no detailed theory of quantified *de re* modal propositions from this period [i.e., the twelfth and thirteenth centuries], and the first attempts in this direction by Abelard and his followers are rather confused. A satisfactory theory of these *de re* modal propositions appears only in the fourteenth century, and the various relations between such propositions was [*sic*] not fully treated until Buridan presents them in his octagon of opposition."

[21] Chatti (2019: 162).

[22] Chatti (2014: 350).

[23] Strobino and Thom (2016: 354–356).

Crockaert P (1512) Summularum artis dialectice vtilis admodum interpretatio super textum magistri Petri Hispani. Per Johannem Cleyn, Lugduni
De Celaya J (1525) Expositio in primum tractatum Summularum Petri Hispani. Venalia prostant in Clauso Brunello, Parrhisiis
Demey L (2019) Boolean considerations on John Buridan's octagons of opposition. Hist Philos Logic 40:116–134
Dolz J (1512) Disceptationes super primum tractatum Summularum. Opera ac caracteribus Johannis de la Roche, Parisius
Dorp J (1495) Summularum liber. Per Johannem Carchagni, [Lyon]
Dorp J (1499a) Perutile compendium totius logice Ioannis Buridani cum preclarissima expositione. Per Petrum Joannem de Quarengiis, Venetiis
Dorp J (1499b) Summula [sic] de dialectica. Per Janonum Carcani, [Lyon]
Dorp J (1504) Summularum liber. Expensis Johannis Granion, Parisius
D'Ors Á (1981) En torno a una figura de oposicion de proposiciones hipoteticas condicional y consecuencia intrinseca. In: Fernández González R (ed) Lógica, epistemologia y teoria de la ciencia. Ministerio de Educación y Ciencia, Madrid, pp 237–250
Dullaert J (1509) Questiones super duos libros Peri hermenias Aristotelis. In edibus Dionisii Roce, Parisius
Dullaert J (1515) Questiones super duos libros Peri hermenias Aristotelis. Impresse per Stephanum Baland, [Lyon]
Dutilh Novaes C (2004) A medieval reformulation of the *de dicto/de re* distinction. In: Běhounek L (ed) The Logica yearbook 2003. Filosofia, Prague, pp 111–124
Enzinas F (1528) Primus tractatus Summularum. Apud Reginaldum Chauldiere, Parisiis
Hubien H (ed) (1976) Iohannis Buridani Tractatus de consequentiis. Publications universitaires, Leuven
Lagerlund H (2000) Modal syllogistics in the middle ages. Brill, Leiden
Major J (1527) Introductorium perutile in Aristotelicam dialecticen duos terminorum tractatus ac quinque libros summularum complectens. In aedibus Ioannis Parvi et Aegidii Gorimontii, [Paris]
Major J (1528) Quaestiones logicales in veterem Aristotelis Dialecticen. Jehan Petit, Parisiis
Noreña CG (1975) Studies in Spanish Renaissance thought. Martinus Nijhoff, Den Haag
Ong WJ (1958) Ramus: method, and the decay of dialogue. From the art of discourse to the art of reason. University of Chicago Press, Chicago
Pardo H (1505) Medulla dyalectices omnes ferme grauiores difficultates logicas acutissime dissoluens. Per Guillermum Anabat, Parisius
Read S (trans) (2015) John Buridan: *Treatise on consequences*. Fordham University Press, New York
Risse W (1998) Bibliographia philosophica vetus. Repertorium generale systematicum operum philosophicorum usque ad annum MDCCC typis impressorum, vol. 2: Logica. Georg Olms, Hildesheim
Soto D (1529) Summule. In officina Joannis Junte, Burgis
Soto D (1554) Summularum aeditio secunda. Excudebat Andreas a Portonariis, Salmanticae
Strobino R, Thom P (2016) The logic of modality. In: Novaes CD, Read S (eds) The Cambridge companion to medieval logic. Cambridge University Press, Cambridge, pp 342–369
Thom P (2003) Medieval modal systems. Problems and concepts. Ashgate, Aldershot
Van der Lecq R (ed) (1983) Johannes Buridanus: *Questiones longe super librum Perihermeneias*. Ingenium, Nijmegen
Van der Lecq R (ed) (2005) Johannes Buridanus: *Summulae de propositionibus*. Brepols, Turnhout

Chapter 8
Conclusion

Abstract The conclusion wraps up the discussion, and summarizes our main claims: (1) the *Wegestreit* shaped the discourse on modal logic during the post-medieval period; (2) Fabri's modal logic implements the approach to the topic that is typical of the Paris *via moderna*; (3) the Traditionalist Thesis about Louvain is in need of revision.

Keywords *Wegestreit* · *Via antiqua* · *Via moderna* · John Fabri of Valenciennes · University of Louvain

This book dealt with developments in modal logic that took place during the *Wegestreit*, a topic that has been almost entirely neglected by past scholarship on the history of logic. The focus of the book was on a lesser known logician named John Fabri of Valenciennes (fl. c. 1500), who was active in Louvain around the turn of the sixteenth century and authored a set of commentaries on the *Organon* that today survive in Saint-Omer, Bibliothèque d'agglomération, MS 609. The book envisaged reconstructing Fabri's modal logic as set out in parts of the commentaries on *De interpretatione* and *Analytica priora*, and to interpret this logic against the backdrop of the divide between the *via antiqua* and *via moderna*, which shaped the debates on logic in Northern Europe at the time. The overall argument of the book was that Fabri is an exponent of the approach to modal logic that is typical of the Paris *via moderna* and is ultimately rooted in the work of John Buridan.

Fabri's modal logic is a token-based logic that is organized around the *sensus compositus-sensus divisus* distinction. In Chap. 2, we showed that the set of modalities in Fabri's logic consists of both alethic and epistemic terms, and we argued that his ascribing modal status to epistemic terms in particular is a first significant indication that his work implements the same approach as the one that is sustained by the Paris *via moderna*. In Chap. 3, we shifted focus to Fabri's account of composite modals, and we laid bare its close ties with his theory of supposition. Fabri, we showed, conceived of *dicta* occurring in composite modals as metalinguistic predicates that supposit materially for occurrences of the assertoric proposition from which they are obtained. A consequence of this view, Fabri maintained that such *dicta* bear quantification. Moreover, we showed that Fabri also endorsed

C. Geudens, L. Demey, *The Modal Logic of John Fabri of Valenciennes (c. 1500)*, SpringerBriefs in Philosophy, https://doi.org/10.1007/978-3-030-98802-9_8

the principle that we dubbed ‘quantification invariance’, which says that every quantifier governing the *dictum* in a composite modal has the same influence on the proposition’s truth conditions (namely, none). After suggesting a model-theoretic reconstruction of Fabri’s logic of composite modals in Chap. 4, we argued in Chap. 5 that the endorsement of both the well-formedness of quantified *dicta* and the principle of quantification invariance is typical of Paris authors who sympathized with the *via moderna*. Finally, in Chap. 6, we discussed Fabri’s logic of divided modals. We discussed Fabri’s procedure of determining the truth value of divided modals, and we showed that by Fabri’s account of the truth conditions of quantified divided modals, the opposition relations between these propositions can be geometrized as an octagon. We pointed out that there are good reasons to believe that Fabri was also aware of this. In Chap. 7, we argued that both his procedure of determining the truth value of divided modals and his incorporating a modal octagon in his logic again connect him with the Paris *via moderna*.

The significance of the book goes well beyond outlining some of the core issues in modal logic during the post-medieval period, however. For one thing, there is a tendency in the scholarship on the history of logic to focus on household names such as Ockham and Buridan. Through a case study of an author whose logic remained entirely unstudied thus far, this book brings us one step closer to achieving the ideal of a history of logic without any gaps. Moreover, as we explained in the introduction (Chap. 1), a common idea in the scholarship on medieval philosophy and intellectual history in general holds that along with Cologne, Louvain was one of the few places in Northern Europe where the *via antiqua* reigned, and where the *via moderna* was almost entirely absent. Some scholars, however, have recently expressed concerns about the truthfulness of this ‘Traditionalist Thesis’. The present book clearly adds to these concerns. Assuming that its main argument is successful and that Fabri’s modal logic does indeed go back to Buridan, the book provides additional evidence that the Traditionalist Thesis is a scholarly construct and in need of revision. Moreover, most of the arguments that have been mounted against the Traditionalist Thesis in the scholarship, are drawn from Louvain texts on topical logic. This book shows that these texts do not constitute an isolated case, and that there is at least one Louvain modal logic that also relies on elements of the logical apparatus that was typical of the *via moderna*. Whether there are more than one is a question for future research to address. To date, we know of eight further commentaries on the *Organon* that originated in Louvain during the fifteenth and early sixteenth centuries:

• Greifswald, Bibliothek des Geistlichen Ministeriums, MS 34 D IX (also preserved in Berlin, Staatsbibliothek – Preußischer Kulturbesitz, MS Magdeburg 227)	c. 1440–1450
• Basel, Universitätsbibliothek, MS F.VII.1	1453
• Cambrai, Bibliothèque municipale, MS 962 (860)	1469
• Edinburgh, university library, MS 205 (Laing MS 149)	1477
• Cambrai, Bibliothèque municipale, MS 964 (862) (also preserved in Saint-Omer, Bibliothèque d’agglomération, MSS 585 and 607)	1481–1482

- Leuven, Universiteitsbibliotheek, MS 237 1513–1514
- Aberdeen, university library, MS 261 After 1450
- Utrecht, Universiteitsbibliotheek, MS 825 After 1450

With the exception of Utrecht, Universiteitsbibliotheek, MS 825, all these texts cover *De interpretatione* and *Analytica priora*, and thus contain passages on modal logic. A study and contextualization of these passages would provide a significant contribution to our knowledge of both the *Wegestreit* and late-medieval (modal) logic, and would help to clarify to what extent Fabri can be considered to have voiced the received view among Louvain logicians of his day.

More generally, still, the book also provided evidence that the *Wegestreit* did indeed shape the discourse on modal logic during the post-medieval period, which has not been sufficiently stressed in the scholarship thus far. The received view among scholars holds that the *Wegestreit* centered on the status of universals, with *antiqui* being reduced to realists and *moderni* to nominalists.[1] If there is one thing this book shows, then it is that this view is too narrow. A branch of logic as particular as modal logic, which has very little to do with universals, was also among the topics on which *antiqui* and *moderni* held very different views. More than anything, this book is a plea to carry out more research into *Wegestreit* philosophy. Even though it is one of the most fascinating areas in the history of philosophy, late-medieval philosophy is a scholarly no-man's land, which is usually described as either the aftermath of medieval philosophy, the antithesis of humanist philosophy or the prelude to Early Modern philosophy, and is rarely considered a domain worthy of specialized study in itself. We need to adopt a different perspective if we ever want to obtain a complete picture of the development of Western philosophy.

References

Klima G (2008) The nominalist semantics of Ockham and Buridan: a 'rational reconstruction'. In: Gabbay DM, Woods J (eds) Handbook of the history of logic, vol. 2: Mediaeval and Renaissance logic. Elsevier, Amsterdam, pp 389–431

Marenbon J (2015) Latin philosophy, 1350-1550. In: Marenbon J (ed) The Oxford handbook of medieval philosophy. Oxford University Press, Oxford, pp 220–244

Normore C (2017) Nominalism. In: Lagerlund H, Hill B (eds) The Routledge companion to sixteenth-century philosophy. Routledge, New York, pp 121–136

[1] See, e.g., Marenbon (2015: 224) for a recent formulation of this view. There are exceptions, of course. Klima (2008) and Normore (2017), for instance, interpret the separation between the *via antiqua* and *via moderna* as the result of a debate that fundamentally centered on semantics rather than ontology.

Appendices

Appendix I: Transcription of Fabri's Questions on *De int.* 12-13

This appendix contains a transcription of Fabri's question commentary on Aristotle's *De int.* 12-13, as reported in **S**, ff. 217vb-222vb. The spelling found in **S** has been retained throughout, though the punctuation has been adapted to modern standards. Becker numbers, denoted '(B)', have been inserted in the margins. The following sigla are used:

<...>	=	*litterae vel verba quae supplenda sunt*
[...]	=	*litterae vel verba quae expungenda sunt*
†...†	=	*lectio incerta*
...	=	*emendatio*
[[...]]	=	*lacuna*
sup. lin.	=	*supra lineam*
corr.	=	*correxit*
add.	=	*addidit*
rep.	=	*repetivit*

The reader may notice that the use of the tenses and moods in the transcription below shows much variation, even according to medieval standards. For instance, there are several occasions where a single clause contains one verb in the indicative mood and another verb in the subjunctive mood; and one and the same verb form occurring in two sentences that are syntactically almost perfectly analogous might have the present tense in one sentence and the imperfect tense in the other. This variation might be due to sloppiness on behalf of **S**'s scribe who, it should be remembered, was only a student, but it might also be a feature of Fabri's own

C. Geudens, L. Demey, *The Modal Logic of John Fabri of Valenciennes (c. 1500)*,
SpringerBriefs in Philosophy, https://doi.org/10.1007/978-3-030-98802-9

Latin style. In absence of conclusive evidence, the transcription gives the verb forms as they appear in **S**. The text has only been emended on those occasions where **S** has problems of subject-verb agreement.

***

<QUESTIONES IOHANNIS FABRI DE VALENCHENIS IN ARISTOTELIS DE INTERPRETATIONE, CAP. XII-XIII>

f. 217vb | HIS AUTEM || Hic philosophus determinat de enunciationibus in quibus aliquid apponitur copule, scilicet de modalibus; et primo determinat de oppositione modalium. 21a34

f. 218ra <i> Notandum primo quod duplices sunt modales. Quedam uocantur modales | composite, ut ille in quibus modus est alterum extremorum, ut subiectum uel predicatum. Alie sunt modales diuise, in quibus *modus*[1] et uerbum situantur inter duo extrema; et illud quod tunc precedit modum et uerbum est subiectum uel tenens se ex parte subiecti, et illud quod sequitur modum et uerbum est predicatum uel tenens se ex parte predicati. Exemplum, ut 'hominem possibile est esse album': hic ly hominem est subiectum et ly album predicatum, et istud aggregatum 'possibile est esse[2]' est copula uel tenens se ex parte copule. Et de istis modalibus diuisis erit hic principaliter sermo.

<ii> Sciendum secundo quod ille modales diuise[3] sunt affirmatiue in quibus nulla ponitur negatio cadens supra copulam uel[4] partem copule, et ille negatiue in quibus ponitur una negatio cadens supra copulam uel partem copule; quod si essent due negationes cadentes supra copulam uel partem copule, tunc essent affirmatiue, et si tres negationes, essent negatiue, et sic consequenter.

<iii> Notandum tertio quod 'necesse' et 'impossibile' dicuntur modi uniuersales, quia sicut se habent 'omnis' et 'nullus' in illis de inesse, ita se habent 'necesse' et 'impossibile' in modalibus. Similiter 'possibile' et 'contingens' sunt modi particulares, quia habent se in modalibus sicut 'quidam' et 'aliquis' habent se in illis de inesse. Ex quo patet quod in modalibus est duplex quantitas, scilicet ex parte modi et ex parte subiecti.

<iv> Notandum quarto circa quantitatem modalium quod ille sunt singulares ex parte subiecti in quibus subicitur terminus singularis, ut 'Sortem possibile est currere', et ille uniuersales in quibus subicitur terminus communis signo uniuersali determinatus. Similiter est de particulari et indiffinita. Sed ex parte modi ille sunt uniuersales que sunt de necessario uel de impossibili, et ille particulares que sunt de possibili uel contingenti. Ex quo patet quod ille sunt simpliciter uniuersales que ex parte utriusque sunt uniuersales, et ille simpliciter particulares que ex parte utriusque

[1] modus] modum **S**

[2] esse] *sup. lin.* **S**

[3] diuise] *corr. ex* affirmatiue **S**

[4] uel] *corr. ex* per **S**

sunt particulares. Sed ille | que ex parte unius sunt uniuersales et ex parte alterius f. 218rb particulares, possunt dici mixte quantitatis. Patet etiam quod omnes de purpurea sunt uniuersales ex parte modi et affirmatiue, et de yliace uniuersales negatiue, de amabimus particulares affirmatiue, et de edentuli negatiue et omnes particulares, et hoc ex parte modi.

<v> Notandum circa oppositiones modalium quod possunt poni tales regule. <a> Prima est quod ille de purpurea et yliace contrariantur, si ambe sunt uniuersales ex parte subiecti uel ad minus una. Probatur, quia tales propositiones sunt eiusdem subiecti et predicati, et una est affirmatiua et alia negatiua, et ambe sunt uniuersales (saltem ex parte modi), et participant legem contrariarum. Istud patet a signo, quia in talibus non est unus terminus qui in neutra distribuatur. Sed si ambe essent particulares ex parte subiecti, tales propositiones essent disparate; hoc est, nullo modo opposite. Quod patet, quia non opponuntur contrarie, quia ibi est aliquis terminus qui nullibi distribuitur; item tales possunt simul esse uere. Neque possunt opponi subcontrarie, quia ibi est aliquid quod utrobique distribuitur, scilicet tempus importatum per copulam; et etiam tales possunt simul esse false, ex quo sequitur quod etiam non sunt contra<dicto>rie. <b> Secunda regula est <quod> propositiones de amabimus et edentuli que sunt eiusdem subiecti et predicati subcontrariantur, si ambe sunt particulares ex parte subiecti uel ad minus una. Sed si ambe essent uniuersales ex parte subiecti, ille sunt disparate. *Idem*[5] dici posset proporcionabiliter probando sicut probatum est in regula precedenti de contrariis. Ex istis regulis sequi uidetur quod propositio de necessario in purpurea uniuersalis ex parte subiecti et propositio uniuersalis ex parte subiecti de possibili in edentuli | f. 218va contrariantur, quia non est ibi aliquis terminus in neutra distributus et ibi est[6] terminus in utraque distributus. Ex illa regula etiam sequitur quod propositio de necessario in purpurea et propositio de possibili in edentuli, si ambe sint particulares ex parte subiecti, subcontrariantur, quia etiam ibi non est aliquis terminus in utraque distributus. [[...]][7] Nec sunt contradictorie, quia ibi est terminus in neutra distributus. Et simili modo dicendum est de aliis, sicut potest uideri in figura de oppositionibus modalium. <c> Tertia regula est quod propositiones de purpurea et edentuli que sunt eiusdem subiecti et predicati contradicunt, si una sit particularis ex parte subiecti et alia uniuersalis. Similiter propositiones de yliace et amabimus que sunt eiusdem subiecti et predicati contradicunt, si una sit uniuersalis ex parte subiecti et alia particularis ex parte subiecti. Istud probatur, quia tales sunt eiusdem subiecti et predicati, et una est affirmatiua et alia negatiua, et sunt diuerse quantitatis tam ex parte modi quam subiecti, et participant legem contradictoriorum. Istud probatur a signo, quia in talibus non est aliquis terminus distribuibilis qui est in utraque

---

[5] idem] item **S**

[6] est] # *add.* **S**

[7] **S** probably contains a lacuna here. In the previous sentence, Fabri said that the particular propositions in *purpurea* and *edentuli* are subcontraries, while in the next sentence he explains why these propositions are not contradictories *either* (note the use of '*nec*' at l. 58). Our scribe might have skipped one or more sentences where Fabri explains why these propositions are not contraries or disparates.

distributus uel qui est in neutra distributus. Sed si tales propositiones essent ambe uniuersales ex parte subiecti, tunc essent contrarie; sed si ambe essent particulares ex parte subiecti, essent subcontrarie, ut predictum est.[8] <d> Quarta regula est <quod> propositiones de purpurea et amabimus eiusdem subiecti et predicati subalternantur, si sit plus de uniuersalitate in una quam in altera. Ex quo sequitur quod uniuersalis ex f. 218vb parte subiecti in purpurea et uniuersalis ex parte subiecti in amabimus | subalternantur, si sint eiusdem subiecti et predicati. Item uniuersalis ex parte subiecti in purpurea et etiam particularis ex parte subiecti ambe subalternantur particulari [s] in amabimus. Particularis in purpurea et uniuersalis in amabimus sunt disparate non habentes aliquam legem oppositionis, et similiter est in iliace et edentuli. <e> Quinta regula est quod modales diuise que sunt singulares ex parte subiecti, quarum una est in purpurea et alia in iliace, eiusdem subiecti et predicati contrariantur; sed si una fuerit in amabimus et alia in edentuli, subcontrariantur, et sic de aliis secundum lineas, sicut dicit Petrus Hispanus.[9]

<i> Arguitur: “Ex prima regula et secunda sequitur quod uni affirmationi essent plures negationes opposite in eodem genere oppositionis non equipollentes, quod est contra philosophum in primo hoc.” Dicitur concedendo sequelam. Ulterius dicitur quod ille non opponuntur tali affirmationi eque directe, sicut dictum est prius in illis de inesse. Istud enim potest contingere in omnibus enunciationibus in quibus est duplex quantitas.

<ii> Arguitur: “Iste sunt contradictorie: ‘omnem hominem necesse est currere’ et ‘aliquem hominem possibile est non currere’; et tamen non *fertur*[10] negatio ante modum, ergo contra textum.” Dicitur quod textus solum uult quod quando datur contradictio de eodem modo, tunc negatio debet ferri ante[11] modum.

<iii> Arguitur: “Iste sunt contradictorie: ‘Sortes currit contingenter’, ‘Sortes non currit contingenter’; et tamen fertur negatio ad uerbum et non ad modum, quod est contra textum.” Dicitur quod licet hic non fertur negatio immediate ad modum, tamen mediate; uel dicitur quod *illius*[12] propositionis potest esse duplex sensus. Unus est iste: ‘Sortes contingenter non currit’; alius est[13] iste: ‘Sortes non f. 219ra contingenter currit’. Secundum primum sensum non contradicit | prime, ut clarum est, sed bene in secundo sensu.

<iv> Arguitur: “Alique sunt modales de uero et falso, de per se et per accidens; de quibus non facit philosophus mentionem, ergo ipse est diminutus.” Dicitur quod auctores communiter non faciunt de illis mentionem; cuius ratio est quia tales

---

[8]The reading ‘*ut predictum est*’ is certain, but Fabri did not point out earlier on that that the particular *iliace* and *amabimus* propositions are subcontrary to each other. So either Fabri made a mistake, or the text is corrupted.

[9]The reference is to Hispanus, *Sum. log.* I.24-25 (De Rijk 1972: 13-16), though Peter nowhere describes or draws a modal octagon, as Fabri is doing here.

[10]fertur] feret **S**

[11]ante] *corr. ex* ad **S**

[12]illius] ille **S**

[13]est] *rep. sed del.* **S**

propositiones non habent equipollentias uel oppositiones ad istas modales de quibus hic agitur. Sed sole modales de istis modis ‘possibile’, ‘contingens’, ‘impossibile’ <et> ‘necesse’ habent equipollentias et oppositiones adinuicem per prepositionem et postpositionem negationum.

<v> Arguitur: “Iste sunt de subiecto uniuersali in purpurea et iliace: ‘omne animal non esse hominem non est possibile’, ‘omne animal esse hominem non est possibile’; et tamen non contrariantur, quia sunt simul uere.” Dicitur quod iste sunt modales composite, de quibus non intelliguntur supradicte regule. Ulterius dicitur quod dicte propositiones non sunt uniuersales ex parte subiecti simpliciter loquendo sed indiffinite, quia in ipsis totum dictum est subiectum; et licet uocentur uniuersales ex parte dicti, tamen illa solum est uniuersalitas secundum quid. Ulterius dicitur quod etiam proprie loquendo non sunt eiusdem subiecti, loquendo scilicet de subiecto tocius propositionis, quia non est idem dictum quod est utrobique subiectum.

<vi> Arguitur: “Ex dictis sequeretur quod modales non essent distinguende penes sensum compositum et diuisum, quia ista modalis ‘album esse nigrum est possibile’ diceretur simpliciter composita, et ista ‘album possibile est esse nigrum’ diuisa diceretur.” | Dicitur[14] concedendo quod de rigore logice tales non sunt distinguende. f. 219rb Sunt tamen secundum usum distincte, communiter scilicet loquentium.

<vii> Arguitur: “Ista est incongrua: ‘Sortem possibile est currere’; ergo uidetur quod propositiones modales in sensu diuiso non sunt *admittende*[15]. Antecedens probatur; primo quia non uidetur a quo regatur ille accusatiuus ‘Sortem’; [quod] secundo quia non uidetur quid est hic suppositum illi uerbo ‘est’; tertio quia non uidetur a quo regatur ly possibile.” Dicitur negando antecedens. Ulterius dicitur quod ille accusatiuus ‘Sortem’ regitur ab illo infinitiuo ‘currere’. Secundo dicitur quod illud dictum ‘Sortem currere’ supplet uicem suppositi respectu illius uerbi ‘est’. Tertio dicitur quod ly possibile regitur †apertius† ab illo uerbo ‘est’ ex ui copule. Sed dicat aliquis: “Iam non habebitur differentia inter istam diuisam et suam compositam, quia utrobique ponuntur eadem regimina grammaticalia.” Dicitur quod immo est differentia, quia si caperetur in sensu composito, tunc dictum supponeret materialiter et supponeret gratia tocius, puta pro propositione categorica preiacente; sed quando intelligitur in sensu diuiso, tunc dictum non supponit gratia tocius, item tunc partes dicti supponunt significatiue, ut ly Sortem pro eo quod est Sortes. Sed dicat aliquis: “Dictum numquam potest supplere uicem suppositi, nisi quando supponit materialiter.” Dicendum quod istud est manifeste falsum; quod patet, quia appellatum potest supplere | uicem suppositi et tamen non suppositione f. 219va materiali, sed *partibus supponentibus*[16] significatiue, ut cum dicitur quod predicatores predicant †nomen† diabolo. Item grammatici dicunt quod quinque modis aliquid supplet uicem suppositi siue nominatiui, †inter quos modos† illi duo modi sunt distincti, quorum unus est quo aliquid supplet uicem nominatiui dum

[14] Dicitur] quod *add. sed del.* **S**

[15] admittende] admittendo **S**

[16] partibus supponentibus] partes supponunt **S**

ponitur materialiter, et alius quo[d] dictum supplet uicem nominatiui. Ergo dictum potest supplere uicem suppositi non supponendo materialiter.

<viii> Petitur hic utrum ista sit negatiua: ‘Sortem possibile est non esse’. Dicitur breuiter quod sic. Probatur, quia ipsa est in edentuli, et nos diximus in regulis quod ille sunt negatiue. Preterea contradictoria eius est affirmatiua, scilicet ‘Sortem necesse est esse’. Iam in categoricis affirmatiua non contradicit affirmatiue. Preterea ipsa equiualet negatiue, isti uidelicet: ‘Sortem non necesse est esse’; ergo est negatiua. Et confirmatur, quia est uera subiecto pro nullo supponente, ergo non est affirmatiua et per consequens est negatiua. <viii-b> Arguitur: “Contradictoria eius est negatiua, ergo ipsa est affirmatiua. Antecedens probatur, quia philosophus dicit quod eius que est ‘possibile est non esse’ opposita negatio est ‘non possibile est non esse’.” Dicitur quod philosophus solum uult illam esse negatiuam de modo, sed
f. 219vb tamen simpliciter est affirmatiua, quia ibi sunt due negationes | cadentes supra copulam uel partem eius. <viii-c> Arguitur: “Copula non negatur, ergo predicta propositio non est negatiua. Antecedens probatur, quia copula est hoc totum ‘possibile est <non> esse’.” Dicitur quod licet toti copule non preponitur negatio, tamen principaliori parti preponitur; et hoc sufficit. Unde principalior pars copule in modalibus diuisis semper est copula que includitur in infinitiuo, quia in ordine ad illam partes in congruitate ordinantur. <viii-d> Replicatur: “Modus est principalis <pars> *copule*[17], ergo contra dicta. Antecedens patet, quia philosophus dicit quod sicut in illis de inesse ‘esse’ et ‘non esse’ sunt appositiones (id est, copule), sic modi in modalibus sunt appositiones.” Dicitur quod ‘sicut’ non dicit ibi omnimodam similitudinem, sed dicit ibi similitudinem in hoc quod sicut in illis de inesse ad dandum contradictoriam negatio debet ferri ad copulam, sic in modalibus debet ferri ad modum. <viii-e> Et si dicatur: “Modales denominantur a modo, sed denominatio fit semper a principaliori parte”; dicitur quod non semper *fit*[18] denominatio a principaliori parte, quia homo denominatur corruptibilis propter naturam, que tamen non est principalior pars in homine. Sequitur textus de equipollen<tiis>.

f. 220ra CONSEQUENTIE || | Hic philosophus determinat de equipollentiis modalium. 22a14

<i> Pro quo nota quod modales diuise de eadem linea equiualent inter se, si sint eiusdem quantitatis ex parte subiecti. Istud probatur, quia quando hoc obseruatur, tunc tales propositiones sunt eiusdem subiecti et predicati, et sunt eiusdem qualitatis, et omnes termini distribuibiles habent se similiter quoad distributionem et non distributionem; ergo tales propositiones sunt adinuicem equiualentes. Preterea, sicut se habent ‘omnis’ et ‘quidam’, ita se habent ‘necesse’ et ‘possibile’; sed ‘omnis’ et ‘quidam’ equiualent per prepositionem et postpositionem negationis; ergo ‘necesse’ et ‘possibile’ equiualent per prepositionem et postpositionem negationis. Item sicut se habent ‘omnis’ et ‘nullus’, ita se habent ‘necesse’ et ‘impossibile’; sed ‘omnis’ et ‘nullus’ equiualent per postpositionem negationis; ergo etiam ‘necesse’ et ‘impossibile’ hoc modo equiualent. Item sicut se habent ‘nullus’ et ‘quidam’, ita se habent ‘impossibile’ et ‘possibile’; sed ‘nullus’ et

[17] copule] copula **S**

[18] fit] sit **S**

'quidam' equiualent per prepositionem negationis; ergo 'impossibile' et 'possibile' sic equiualent. Iam ista hoc modo obseruantur in lineis quando propositiones sunt eiusdem subiecti quoad quantitatem, ergo et cetera.

<i-a> Arguitur: "Secundum Petrum Hispanum quarta propositio que est de necessario semper debet esse de dicto contradictorio, sicut patet in regulis;[19] ergo contra dicta." Dicitur[20] quod illud dictum Petri Hispani quoad propositionem de necessario habet tantummodo ueritatem de compositis, de quibus adhuc nihil diximus. <i-b> Replicatur: "Ex opposito antecedentis non semper sequitur oppositum consequentis; et licet sequeretur, tamen non sequeretur tantummodo contradictorium consequentis. Et tamen ista uidetur philosophus presupponere in textu, ergo rationes philosophi non sunt efficaces." Dicitur concedendo totum antecedens. Ulterius dicitur | quod in equipollentibus talia bene presupponuntur, ut f. 220rb si ad antecedens sequatur consequens per equipollentiam, tunc ad contradictorium antecedentis sequitur tantummodo contradictorium consequentis per equipollentiam, quia nulla alia propositio que non esset contradictoria consequentis posset in tali casu equiualere contradictorie antecedentis, ut patet intuenti.

<ii> Arguitur: "Iste sunt in eadem linea: 'hominem non possibile est currere', 'hominem impossibile est currere'; et non equipollent, ergo propositiones eiusdem linee non semper sunt equiualentes. Secunda pars antecedentis probatur, quia prima est negatiua et secunda affirmatiua." Dicitur negando quod secunda est affirmatiua; et licet in ea non ponatur negatio 'non' explicite, tamen implicite, quia 'impossibile' est modus negatiuus.

<iii> Arguitur: "Iste sunt in eadem linea: 'hominem contingit currere' et 'hominem non necesse est non currere'; sed iste non equiualent, ergo idem quod prius. Minor patet, quia ad primam aliquid sequitur quod non sequitur ad secundam; ergo non equiualent. Antecedens probatur, quia ad primam sequitur ista: 'homo possibiliter non currit'; que non sequitur ad secundam. Dicitur quod 'contingens' capitur dupliciter. Uno modo proprie et specialiter, ut tantum ualet sicut 'non necessarium nec impossibile'; alio modo generaliter, ut tantum ualet sicut 'non impossibile'. Primo modo capiendo procedit argumentum, sed ita non hic sumitur, quia propositiones de contingenti sic capto non equiualent aliis; sed 'contingens' hic capitur secundo modo, hoc est generaliter.

<iv> Arguitur sic: "Iste sunt in eadem linea: 'hominem necesse est currere' et 'hominem non possibile | est non currere'; sed iste non equipollent, ergo idem quod f. 220va prius. Minor probatur, quia propositiones equiualentes debent habere eandem copulam. Iam sic non est hic, quia copula prime est illud aggregatum 'necesse est esse', et copula secunde est illud aggregatum 'non possibile est non esse'. Etiam propositiones equiualentes debent subordinari eidem mentali, sed iste non subordina<n>tur eidem mentali, quia in mentali prime non est aliqua negatio et in ea ponitur ille modus 'necesse', et in mentali secunde sunt due negationes et in ea ponitur 'possibile'." Propter istud argumentum est opinio Marcilii quod in omnibus

[19] See Hispanus, *Sum. log.* I.24 (De Rijk 1972: 13-14).

[20] Dicitur] *sup. lin.* **S**

modalibus ille modus 'possibile' est copula uel pars copule.[21] Diceret ergo quod etiam in proposito in mentali prime est iste modus 'possibile' cum duabus negationibus. Unde iste uult quod propositiones equiualentes debent subordinari eidem mentali. Sed ista opinio non est probabilis. Respondetur ergo ad argumentum concedendo quod in mentali unius ponuntur due negationes et in mentali alterius nulla; item quod in mentali unius ponitur ille modus 'necesse' et in mentali alterius ille modus 'possibile'. Ulterius dicitur quod argumentum bene probat quod dicte propositiones non equiualent in significando quasi essent sinonime. Tamen equiualent in ueritate et falsitate formaliter, sicut prius dictum est in primo;[22] et taliter debent propositiones equiualentes se habere, et non aliter.

<v> Dubitatur <v-a> utrum modales composite sunt uere modales; <v-b> secundo quomodo[23] attendatur proprie quantitas in ipsis; <v-c> tertio quomodo in
f. 220vb ipsis proprie attenditur qualitas; <v-d> quarto quomodo in | ipsis attenditur oppositio; <v-e> quinto dubium habetur quomodo proprie attenditur in ipsis equiualentia. <v-a> Ad primum dicitur quod propositiones modales composite non sunt uere modales. Probatur, quia ille non sunt modales in [in] quibus copula non modificatur, ergo modales composite non sunt uere modales. Minor probatur, ut in ista 'album esse nigrum est impossibile' ly est, quod est copula, non modificatur. <v-b> Ad secundum dicitur quod quantitas attenditur proprie in modalibus compositis sicut in propositionibus de inesse. Illud probatur, quia modales composite sunt uere de inesse, ut iam dictum est. Ex quo patet quod illa est proprie uniuersalis in qua subicitur terminus communis signo uniuersali determinatus, ut 'omne[24] hominem currere est possibile'. Hic dictum (scilicet 'hominem currere') est terminus communis supponens materialiter et determinatur signo uniuersali. Illa uero est indiffinita in qua subicitur dictum sine[25] signo, ut ista 'hominem currere est possibile'; et illa est singularis in qua subicitur dictum cum *pronomine*[26] demonstratiuo primitiue speciei, ut 'hoc hominem currere est possibile'. <v-c> Ad tercium dicitur quod in modalibus compositis proprie attenditur qualitas sicut in illis de inesse. Ex quo sequitur quod ille sunt affirmatiue in quibus predicatum affirmatur de subiecto, ut ista 'hominem currere est possibile'; et ille sunt negatiue in quibus predicatum negatur a subiecto, ut 'hominem currere non est possibile'. <v-d> Ad

[21] The Marsilius in question is Marsilius of Inghen, though it is unclear which passage Fabri has in mind. There is no evidence that Marsilius's writings were circulating in Louvain around the turn of the sixteenth century; while his account of the copula is summarized in several fifteenth-century texts, including Peter Crockaert's commentary on the *Summulae*, Nicholas of Amsterdam's commentary on the *logica vetus*, and John Dorp's commentary on Buridan's *Summulae de dialectica*; see de Amsterdammis, *in De int.* (Bos 2016: 342 [ll. 9-26]); Crockaert, *in Sum. log.* (Crockaert 1512: sig. d v^{ra}); Dorp, *Perutile compendium* (Dorp 1499: sig. d 2^{rb-va}). Fabri presumably had second-hand knowledge of Marsilius' views, through such a later testimony.

[22] See above, **S**, ll. 170–185.

[23] quomodo] *corr. ex* ut **S**

[24] omne] omnem *add. sed del.* **S**

[25] sine] *sup. lin.* **S**

[26] pronomine] propositione **S**

quartum <dicitur quod> oppositio[27] attenditur proprie in modalibus compositis sicut in illis de inesse. Ex quo sequitur quod uniuersalis affirmatiua et uniuersalis negatiua eiusdem subiecti et predicati sunt contrarie, ut ‘omne hominem currere est possibile’ <et> ‘nullum hominem currere est possibile’, capiendo ly nullum in neutro genere et ly hominem currere pro qualibet propositione preiacente[28] cuius est dictum. Et omnino eodem modo considerandum est de aliis oppositionibus. | f. 221ra <v-e> Ad quintum dicitur quod equiualentia proprie attenditur in modalibus compositis sicut in illis de inesse. Ex quo sequitur quod contrarie equiualent per postpositionem negationis et contradictorie per prepositionem et cetera. Exemplum, ut iste equiualent: ‘non omne hominem currere est possibile’, ‘aliquod hominem currere non est possibile’.

<vi-a> Dubitatur quomodo attenduntur improprie quantitas et qualitas in modalibus compositis. <vi-b> Dubitatur ulterius utrum modales composite que sunt in eadem linea proprie equiualent. <vi-c> Tercio dubitatur utrum ille que sunt in contrariis lineis proprie contrariantur, et que sunt in contradictoriis lineis proprie contradicant, et sic de aliis. <vi-a> Ad primum dicitur quod attribuendo modalibus compositis quantitatem et qualitatem improprie et similitudinarie, pro quanto uidelicet similitudinem habent cum diuisis, ille sunt uniuersales in quibus subiectum dicti sumitur uniuersaliter, et ille particulares que sunt de subiecto dicti particulariter sumpto, et sic de aliis, ut ista esset uniuersalis: ‘omnem hominem currere est possibile’; et ista particularis: ‘aliquem hominem currere est possibile’. Item ille dicuntur affirmatiue que sunt de dicto affirmatiuo, et ille negatiue que sunt de dicto negatiuo, ut ista dicitur[29] negatiua: ‘hominem non currere est possibile’. <vi-b> Ad secundum dicitur quod non, quia oportet propositiones equiualentes et propositiones oppositas esse eiusdem subiecti et predicati, sed modales composite eiusdem linee non sunt eiusdem subiecti et predicati proprie loquendo, ut patet ex precedentibus; ergo non sunt proprie equiualentes. <vi-c> Preterea modales composite que sunt in contrariis lineis, uel que sunt in lineis subcontrariis, et sic de aliis, | f. 221rb non sunt eiusdem subiecti et predicati, ergo non sunt proprie opposite. Propositiones uero que sunt in contradictoriis lineis aliquando sunt eiusdem subiecti et predicati, et tamen non sunt contradictorie nisi una sit uniuersalis et alia particularis illo modo sicut prius dictum est in primo dubio.[30]

<vii-a> Dubitatur circa solutionem istius secunde dubitationis[31] utrum modales composite de eadem linea sequuntur ad se inuicem, licet non proprie equipolleant. <vii-b> Ulterius dubitatur an modales composite in oppositis lineis existentes participent legem propositionum oppositarum, licet non proprie opponantur. <vii-a> <Ad primum dubium dicendum quod sic,> et hoc oportet obseruare quod quarta propositio que est de necessario semper sit opposite quantitatis in ordine ad alias quoad dictum. Item quando arguitur a negatiuis ad affirmatiuas, tunc debet poni

[27] oppositio] *corr. ex* propositio **S**

[28] preiacente] preiacende **S**

[29] dicitur] tunc *add. sed del.* **S**

[30] See ll. **S**, ll. 254–260, or perhaps rather ll. **S**, ll. 60–70.

[31] dubitationis] ulterius *add. sed del.* **S**

constantia propositionis pro qua supponit dictum consequentis, uerbi gratia sic arguendo: 'hominem esse asinum non est possibile, et talis propositio est: 'omnis homo non est asinus'; ergo omnem hominem non esse asinum est necessarium'. Et ratio istius est, quia si talis propositio non esset, tunc non deuocaretur neque uera, neque falsa, neque necessaria. Et fundantur ista in tribus regulis. <a> Prima est: si aliqua propositio est possibilis, ipsa non est impossibilis. Ista regula est manifesta, quia 'possibile' et 'impossibile' sunt disparata, ergo non possunt eidem conuenire. <b> Secunda regula est: si aliqua propositio est possibilis, sua contradictoria non est necessaria. Ista regula etiam patet, quia si sua contradictoria esset necessaria, tunc ipsa semper esset uera; et si tunc altera esset possibilis, tunc ipsa posset esse uera, et sic contradictorie essent simul uere. <c> Tertia regula est: si aliqua propositio est impossibilis, sua contradictoria est necessaria. Ista regula patet, quia si sua contradictoria non esset necessaria | f. 221va sed posset esse falsa et cum illa ponatur impossibilis (et ita immutabiliter falsa), tunc contradictorie possent simul esse false. <vii-b> Ad secundum dubium dicendum quod sic, seruatis quibusdam regulis. <a> Prima regula est quod propositio de necessario in purpurea et propositio de necessario in iliace ad hoc[32] <quod> participent legem contrariarum, debent ambe esse de dicto uniuersali uel ad minus una. De propositionibus de aliis modis que sunt in dictis lineis dicendum est secundum quod istis de necessario equiualent. Ex quo patet quod ad hoc quod propositiones de aliis modis in dictis lineis contrarientur, oportet quod ambe sint de dicto particulari uel ad minus una; et ad hoc quod propositio de necessario in una[33] linea habeat contrarietatem cum propositione de alio modo, oportet quod si ipsa sit de dicto particulari, quod altera etiam sit de dicto particulari. <b> Secunda regula est quod ad hoc quod propositiones de possibili, de contingenti et de impossibili que sunt in amabimus et edentuli subcontrarientur, oportet quod ambe sint de dicto particulari uel ad minus una; si uero ambe sint de necessario, oportet quod ambe sint de dicto uniuersali uel ad minus una; si uero una sit de necessario et alia de aliquo alio modo, oportet quod si illa de alio modo sit de dicto uniuersali, quod illa de necessario sit de dicto uniuersali. <c> Tertia regula est <quod> ad hoc quod propositiones composite que sunt in contradictoriis lineis participent legem contradictoriarum, oportet semper quod dicta sunt eiusdem quantitatis uno casu excepto, scilicet quando una propositio est de necessario et alia de aliquo alio modo. Tunc enim oportet obseruare quod una sit de dicto uniuersali et alia de dicto particulari. <d> Quarta regula est quod ad hoc quod propositio de necessario in purpurea | f. 221vb et illa de possibili in amabimus habeant legem subalternarum, oportet quod si illa de possibili sit de dicto uniuersali, quod tunc etiam illa de necessario in purpurea sit de dicto uniuersali. De aliis autem propositionibus, que scilicet sunt de aliis modis, considerandum est secundum quod istis equipollent, et similiter est in purpurea et edentuli. Sequitur textus.

---

[32] ad hoc] *sup. lin.* **S**

[33] una] *corr. ex* prima **S**

2b29 (B)DUBITABIT || In hoc tertio capitulo philosophus mouet istam questionem: utrum ad 'necesse' sequitur 'possibile'. Hoc enim in presenti textu presuppositum est; et terminat philosophus quod sic, capiendo 'possibile' generaliter.

<i> Notandum primo quod isti quattuor termini 'possibile', 'contingens', 'impossibile' et 'necesse' aliquando capiuntur pure sincathegorematice et aliquando cathegorematice. Quando capiuntur primo modo, tunc oportet imaginari quod sicut ly est captum sincategorematice seorsum nihil significat, et quando ponitur inter duo extrema, tunc significat quod predicatum conuenit subiecto; ita isti quatuor termini seorsum nihil significant, sed quando ponuntur inter duo extrema *iuxta*[34] copulam 'est', tunc significant quod predicatum tali uel tali modo conuenit subiecto, puta necessario uel possibiliter. Et sic patet quod hoc modo capiendo istos terminos ipsi sunt sincathegorematici.

<ii> Notandum secundo quod quando capiuntur cathegorematice, tunc quilibet *eorum*[35] per se potest esse subiectum propositionis uel predicatum, et tunc non ponuntur iuxta copulam, et tunc ly possibile significat tantum sicut istud complexum: 'propositio que qualitercumque significat esse uel non esse, ita possibile est esse uel non esse'. Verbi gratia, cum dico de aliqua propositione quod ipsa est possibilis, ego uolo dicere quod ipsa est propositio que qualitercumque significat esse uel non esse, ita possibile est esse uel non esse. Et sic est simili modo dicendum de aliis tribus terminis.

<iii> Notandum tertio quod capiendo 'necesse' et 'possibile' primo modo quando dicitur "ad 'necesse' sequitur 'possibile'", est directe ac si diceretur in simili "ad 'omnis' sequitur 'quidam'"; et quia tunc sensus est quod ad propositionem | in qua f. 222ra
ponitur 'omnis' sequitur propositio in qua ponitur 'quidam', sic in proposito sensus est quod ad 'necesse' sequitur 'possibile', id est, ad[36] propositionem in qua ponitur illud sincategorema 'necesse' ad copulam sequitur propositio in qua ponitur istud sincategorema 'possibile' ad copulam, ut bene sequitur 'Deus necessario est ens, ergo Deus possibiliter est ens', capiendo 'possibile' generaliter, sicut in simili sequitur 'omnis homo est animal, ergo quidam homo est animal'.

<iv> Notandum quarto quod capiendo istos terminos 'necesse' et 'possibile' secundo modo, quando tunc dicitur quod ad 'necesse' sequitur 'possibile', tunc est directe ac si diceretur in alio simili quod ad hominem sequitur animal per locum a specie ad genus, ut sicut bene ualet 'hec res est homo, ergo hec res est animal', ita per eundem locum bene ualet 'hec propositio est necessaria, ergo hec propositio est possibilis', et hoc capiendo 'possibile' generaliter. Ex quo patet quod capiendo dictos terminos secundo modo, tunc habent se proprie sicut magis commune et minus commune; sed capiendo primo modo, tunc esset omnino improprie dictum quod haberent se sicut magis commune et minus commune, sicut in simili esset improprie dictum quod 'omnis' et 'quidam' haberent se sicut magis commune et minus.

[34] iuxta] inter **S**

[35] eorum] earum **S**

[36] possibile id est ad] *rep. sed del.* **S**

<v> Notandum quod 'contingens' uel 'possibile' specialiter captum ualet tantum sicut 'nec necessarium nec impossibile'; et potest etiam capi dupliciter. Uno modo sincathegorematice, et tunc ponitur ad copulam, ut 'Sortes contingenter est currens', cuius sensus est 'Sortes nec necessario nec impossibiliter est currens'. Alio modo cathegorematice, et tunc ponitur unum extremum propositionis, ut cum dico (demonstrata aliqua propositione) "ista propositio est contingens". Unde sensus est: 'ista propositio nec est necessaria, nec impossibilis'; et istud 'contingens' non sequitur ad 'necesse'.

<vi> Notandum quod non potest bene declarari pro simili quomodo iste terminus f. 222rb 'contingens' *captus*[37] sincathegorematice se habet ad 'necesse' cap|tum sincathegorematice. Si tamen ly quidam ualeret tantum sicut 'aliquis' precise, ita quod 'non omnis', tunc sicut 'quidam' se haberet ad 'omnis', ita se haberet 'contingens' ad 'necesse'; et sicut tunc non sequeretur 'omnis homo est animal, ergo quidam homo est animal', ita nunc non sequitur 'Deum necesse est esse, ergo Deum contingens est esse'.

<vii> Notandum ultimo quod capiendo 'contingens' specialiter et cathegorematice, tunc habet se ad 'necesse' et etiam ad 'impossibile' cathegorematice capta sicut disparatum, nam de nullis eisdem propositionibus possunt uerificari. Sed 'contingens' et 'possibile' generaliter *capta*[38] habent se sicut magis commune et minus, nam omnis propositio contingens est propositio possibilis, sed non econtra.

<i> Arguitur: "Nos dicimus quod Antichristus est possibilis et quod omnis creatura est contingens et quod Deus est necessarius et simera impossibilis, sed isti termini non capiuntur hic sicut prius dictum est, ergo uidetur quod possunt adhuc plu<ribus> modis capi." Dicitur concedendo istud, quia quando capiuntur cathegorematice, tunc possunt capi dupliciter: uno modo ut sunt differentie propositionum, et sic cepimus prius; alio modo ut sunt differentie entium, et sic capiuntur in istis propositionibus.

<ii> Arguitur: "Bene sequitur: 'possibile' specialiter captum non sequitur ad 'necesse', ergo 'possibile' non sequitur ad 'necesse'; ab inferiori ad superius." Dicitur quod in illo loco non debet argui terminis supponentibus materialiter uel simpliciter. Iam hic termini supponunt materialiter.

<iii> Arguitur: "Omnis potentia est ducibilis ad actum, ergo male dicitur in textu[39] quod quedam sunt semper in potentia." Dicitur quod omnis potentia ducibilis est ad actum uel perfectum uel imperfectum. Est tamen aliqua potentia que non potest deduci ad actum perfectum, ut potentia qua continuum dicitur diuisibile in infinitum.

<iv> Dubitatur utrum sole potentie rationales sint ad opposita. Dicitur primo quod sola uoluntas est proprie potentia rationalis, quia ipsa primo obedit rationi et alie f. 222va potentie obediunt uoluntati. Secundo dicitur quod | sola uoluntas proprie est ad

---

[37] captum] capitur **S**

[38] capta] captum **S**

[39] textu] *corr. ex* tractatu **S**

opposita ex se, quia ipsa sola omnibus extrinsecis eodem modo se habentibus potest in quodlibet oppositorum seorsum. Tertio dicitur quod multe alie potentie sunt ad opposita, sed non ex se sed propter uariationem illorum que sunt extra. Verbi gratia, ignis aliquando comburit, scilicet *quando*[40] sibi applicatur combustibile, et sic non potest non[41] comburere quantum est de se; et aliquando non comburit, scilicet quando non habet combustibile sibi applicatum. Item sol aliquando indurat et aliquando mollificat, sed in ista opposita non potest ex se et omnibus extrinsecis eodem modo se habentibus, sed hoc est propter diuersitatem materie circa quam agit sol, ut ceram mollificat et lutum indurat. Similiter canis aliquando prosequitur leporem et aliquando non, tamen non habet ex se quod posset in illa opposita, sed propter uariationem eorum que sunt extra, ut si non impeditur et lepus presentetur tamquam aliquid potens apprehendi a cane, et tunc non est in potestate canis non prosequi. Et sic potest sustineri quod sola uoluntas uel aliquid habens uoluntatem potest ex se omnibus aliis exterioribus eodem modo se habentibus uelle aliquid agere uel non agere.

<v> Dubitatur secundo quomodo cognoscitur de omnibus modalibus an sint uere uel false. Dicitur quod hoc cognoscitur per positionem inesse, hoc est, reducendo eas ad propositiones de presenti et de inesse; de quo dantur alique regule. <a> Prima regula est quod modales composite ponuntur inesse mutando accusatiuum dicti in nominatiuum, et infinitiuum in uerbum personale presentis temporis indicatiui modi, ut ‘album esse nigrum est possibile’ ponitur sic inesse: ‘album est nigrum’. Et tunc oportet uidere utrum modus positus in modali possit uere predicari de tali propositione posita inesse; et si sic, tunc modalis fuit uera, si non, est falsa. | <b> f. 222vb
Secunda regula est <quod> modales diuise que sunt de subiecto singulari non cognotatiuo ponuntur inesse sicut iam dictum est de modalibus compositis, ut ‘Sortem possibile est currere’ ponitur inesse per istam: ‘Sortes currit’. <c> Tertia regula est quod modales diuise que sunt de termino singulari cognotatiuo ponuntur inesse capiendo pronomen demonstratiuum et demonstrando illud pro quo supponebat subiectum ipsius modalis, ut ‘istum sedentem possibile est ambulare’ ponitur sic inesse: ‘iste ambulat’, dimittendo illud cognotatiuum. Et tunc oportet uidere utrum modus positus in modali possit uerificari de tali propositione de inesse; si sic, tunc modalis fuit uera, si non, falsa. <d> Quarta regula est quod modalis diuisa particularis uel indiffinita ponitur inesse per multas singulares in quibus per pronomina demonstratiua demonstrantur ille res pro quibus supponebat subiectum in ipsa modali, ut ‘album possibile est esse nigrum’ ponitur sic inesse: ‘hoc est nigrum’, ‘illud est nigrum’, et sic consequenter demonstrando quodlibet pro quo supponebat ‘album’, et tunc oportet uidere utrum modus positus in modali competat alicui tali propositioni de inesse; quod si sic, modalis erat uera, si non, falsa. <e> Quinta regula est <quod> modales diuise uniuersales ex parte subiecti ponuntur inesse sicut iam dictum est in regula precedente. Sed postquam posite sunt inesse, tunc oportet uidere utrum modus positus in modali possit uere predicari de qualibet

[40] quando] quod **S**
[41] non] *sup. lin.* **S**

tali propositione de inesse; quod si sic, tunc fuit uera, si non, fuit falsa. Ex quo patet quod ista est uera: 'qualibet forma materia potest informari'; et ista falsa: 'materia potest informari qualibet forma'.

<FINIUNT QUESTIONES IOHANNIS FABRI DE VALENCHENIS IN ARISTOTELIS DE INTERPRETATIONE, CAP. XII-XIII>

References

Bos EP (ed) (2016) Nicholas of amsterdam: Commentary on the *Old Logic*. John Benjamins, Amsterdam

Crockaert P (1512) Summularum artis dialectice vtilis admodum interpretatio super textum magistri Petri Hispani. Per Johannem Cleyn, Lugduni

De Rijk LM (ed) (1972) Peter of Spain (Petrus Hispanus Portugalensis): *Tractatus*, called afterwards *Summule logicales*. First critical edition from the manuscripts. Van Gorcum, Assen

Dorp J (1499) Perutile compendium totius logice Ioannis Buridani cum preclarissima expositione. Per Petrum Joannem de Quarengiis, Venetiis

Appendix II: Formal Proofs

This appendix should be read in connection with Chap. 4. In this chapter, two main modal systems were introduced: *ML*, which has (PT) among its axioms and (ERed) among its theorems, and *ML'*, which has (PT') and (ERed) among its axioms. It might be convenient to briefly recall the meaning of these principles, as they repeatedly occur in the proofs below:

- (PT): $\mathbb{M}, t \nvDash (E\varphi \rightarrow T\varphi) \wedge (E(\neg\varphi) \rightarrow T(\neg\varphi))$, for every $\mathbb{M}, t$
- (PT'): $\mathbb{M}, t \vDash (E\varphi \wedge E(\neg\varphi)) \rightarrow \neg(T\varphi \wedge T(\neg\varphi))$, for every $\mathbb{M}, t$
- (ERed): For every $\mathbb{M}$, t and φ: $\mathbb{M}, t \vDash E\varphi$

The appendix gives proofs of the universal validity of the implications (a')-(e') in these systems. For each implication it is possible to formulate a proof that only relies on principles that are valid in both *ML* and *ML'*, and, thus, holds for both systems. Notice that (ERed), which is used in each of the proofs below, except for that of (a'), occupies a slightly special place among these principles. (ERed) is valid in both *ML* and *ML'*, but it has a different status in each system: it is a theorem in *ML* and an axiom in *ML'*. The principle should be interpreted accordingly depending on whether the proofs are taken to be in *ML* or in *ML'*.

$$\text{(a')}\quad \mathbb{M}, t \vDash T(\Box(\underline{\psi})) \rightarrow T\psi, \text{ for every } \mathbb{M}, t$$

Proof Consider an arbitrary time t in a model $\mathbb{M}$, and assume that $\mathbb{M}, t \vDash T(\Box(\underline{\psi}))$. By ($T$) and (TC), we have that $\mathbb{M}, t \vDash E\psi$. To see that $\mathbb{M}, t \vDash T\psi$, notice that from $\mathbb{M}, t \vDash T(\Box(\underline{\psi}))$, it follows by ($T$) that $\mathbb{M}, t \vDash E(\Box(\underline{\psi})) \wedge \Box(\underline{\psi})$, which in turn implies that $\mathbb{M}, t \vDash \Box(\underline{\psi})$ by ($\wedge$). But, by ($\Box$), $\mathbb{M}, t \vDash \Box(\underline{\psi})$ just in case $\mathbb{M}, t \vDash E\psi$, which was already available anyway, and $\mathbb{M}, t' \vDash E\psi \rightarrow T\psi$, for all t' such that tRt'. Thus, $\mathbb{M}, t \vDash E\psi \rightarrow T\psi$ by the reflexivity of R; and since $\mathbb{M}, t \vDash E\psi$, we also have that $\mathbb{M}, t \vDash T\psi$, by *modus ponens*. ∎

$$\text{(b')}\quad \mathbb{M}, t \vDash T(\Box(\underline{\psi})) \rightarrow T(\Diamond(\underline{\psi})), \text{ for every } \mathbb{M}, t$$

Proof Consider an arbitrary time t in a model $\mathbb{M}$, and assume that $\mathbb{M}, t \vDash T(\Box(\underline{\psi}))$. We show that $\mathbb{M}, t \vDash T(\Diamond(\underline{\psi}))$. Notice, first, that $\mathbb{M}, t \vDash T(\Box(\underline{\psi}))$ implies that $\mathbb{M}, t \vDash T\psi$, by (a'), which in turn implies that $\mathbb{M}, t \vDash E\psi \rightarrow T\psi$, by ($\rightarrow$). By the reflexivity of R, it follows that for some t' such that tRt', $\mathbb{M}, t' \vDash E\psi \rightarrow T\psi$, viz. t itself – call this

(A). Moreover, we also have that $\mathbb{M}, t \vDash E\psi$, by (ERed). But (A) along with $\mathbb{M}, t \vDash E\psi$ means that $\mathbb{M}, t \vDash \Diamond(\underline{\psi})$, by ($\Diamond$). Notice, second, that $\mathbb{M}, t \vDash E(\Diamond(\underline{\psi}))$ by (ERed). But $\mathbb{M}, t \vDash \Diamond(\underline{\psi})$ and $\mathbb{M}, t \vDash E(\Diamond(\underline{\psi}))$ imply $\mathbb{M}, t \vDash E(\Diamond(\underline{\psi})) \wedge \Diamond(\underline{\psi})$, by ($\wedge$). By ($T$), this is equivalent to $\mathbb{M}, t \vDash T(\Diamond(\underline{\psi}))$, which finishes the proof. ■

$$\text{(c')} \quad \mathbb{M}, t \vDash T(\Box(\underline{\psi})) \rightarrow T(\neg\Diamond(\neg\psi)), \text{ for every } \mathbb{M}, t$$

Proof Consider an arbitrary time t in a model $\mathbb{M}$, and assume that $\mathbb{M}, t \vDash T(\Box(\underline{\psi}))$. We show that $\mathbb{M}, t \vDash T(\neg\Diamond(\neg\psi))$. As a first step, notice that since $\mathbb{M}, t \vDash T(\Box(\underline{\psi}))$, it is also the case that $\mathbb{M}, t' \vDash E\psi \rightarrow T\psi$, for all t' such that tRt', by (T), ($\wedge$), and ($\Box$) – call this (A). Now consider the following claim.

$$\text{(B)} \quad \text{For every } t' \text{ such that } tRt', \mathbb{M}, t' \vDash E(\neg\psi) \wedge \neg T(\neg\psi)$$

Proof of (B) Let t' be an arbitrary time such that tRt'. By (A), we obtain that $\mathbb{M}, t' \vDash E\psi \rightarrow T\psi$, and by (ERed), we obtain that $\mathbb{M}, t' \vDash E\psi$. Thus also $\mathbb{M}, t' \vDash T\psi$, by *modus ponens*, and $\mathbb{M}, t' \vDash \psi$, by ($T$) and ($\wedge$). But this implies that $\mathbb{M}, t' \nvDash \neg\psi$, by ($\neg$), which in turn implies that $\mathbb{M}, t' \vDash \neg T(\neg\psi)$, by ($T$) and ($\neg$). Since $\mathbb{M}, t' \vDash E(\neg\psi)$ by (ERed), we obtain that $\mathbb{M}, t' \vDash E(\neg\psi) \wedge \neg T(\neg\psi)$ by ($\wedge$), as required. ■

By ($\wedge$) and ($\rightarrow$), we obtain from (B) that there is no t' such that tRt' and $\mathbb{M}, t' \vDash E(\neg\psi) \rightarrow T(\neg\psi)$. This implies that $\mathbb{M}, t \vDash \neg\Diamond(\neg\psi)$, by ($\Diamond$). But $\mathbb{M}, t \vDash E(\neg\Diamond(\neg\psi))$, by (ERed), and thus also $\mathbb{M}, t \vDash E(\neg\Diamond(\neg\psi)) \wedge \neg\Diamond(\neg\psi)$, by ($\wedge$). By ($T$), this is equivalent with $\mathbb{M}, t \vDash T(\neg\Diamond(\neg\psi))$, which finishes the proof. ■

$$\text{(c*')} \quad \mathbb{M}, t \vDash T(\neg\Diamond(\neg\psi)) \rightarrow T(\Box(\underline{\psi})), \text{ for every } \mathbb{M}, t$$

Proof Consider an arbitrary time t in a model $\mathbb{M}$, and assume, towards a contradiction, that $\mathbb{M}, t \vDash T(\neg\Diamond(\neg\psi))$ and yet $\mathbb{M}, t \vDash \neg T(\Box(\underline{\psi}))$. Notice, first, that since $\mathbb{M}, t \vDash T(\neg\Diamond(\neg\psi))$, it is also the case that $\mathbb{M}, t \vDash \neg\Diamond(\neg\psi)$, by ($T$) and ($\wedge$). By ($\Diamond$) and De Morgan, this means that either (A) $\mathbb{M}, t \nvDash E(\neg\psi)$ or (B) there is no t' such that tRt' and $\mathbb{M}, t' \vDash E(\neg\psi) \rightarrow T(\neg\psi)$. But $\mathbb{M}, t \vDash E(\neg\psi)$, by (ERed). Thus, (A) is false, and (B) is true. But (B) is true just in case for all t' such that tRt', $\mathbb{M}, t' \nvDash E(\neg\psi) \rightarrow T(\neg\psi)$. By ($\rightarrow$), this implies that $\mathbb{M}, t' \nvDash T(\neg\psi)$, for all t' such that tRt', meaning there is no t' such that tRt' and $\mathbb{M}, t' \vDash T(\neg\psi)$. Notice, second, that by (T) and De Morgan $\mathbb{M}, t \vDash \neg T(\Box(\underline{\psi}))$ just in case $\mathbb{M}, t \vDash \neg E(\Box(\underline{\psi})) \vee \neg\Box(\underline{\psi})$, which reduces to $\mathbb{M}, t \vDash \neg\Box(\underline{\psi})$ since $\mathbb{M}, t \vDash E(\Box(\underline{\psi}))$ by (ERed). By ($\Box$) and De Morgan, this implies that either (C) $\mathbb{M}, t \vDash \neg E\psi$ or (D) there is a t' such that tRt' and $\mathbb{M}, t' \nvDash E\psi \rightarrow T\psi$.

Since $\mathbb{M}, t \vDash E\psi$ by (ERed), (C) is false, so (D) is true. By ($\rightarrow$), we obtain from (D) that there is a t' such that tRt' and $\mathbb{M}, t' \nvDash T\psi$. By ($T$), ($\neg$) and De Morgan, this implies that there is a t' such that tRt' and $\mathbb{M}, t' \vDash \neg E\psi \vee \neg\psi$, which reduces to $\mathbb{M}, t' \vDash \neg\psi$ since $\mathbb{M}, t' \vDash E\psi$ by (ERed). Applying (ERed) once again, along with ($\wedge$), we obtain that there is a t' such that tRt' and $\mathbb{M}, t' \vDash E(\neg\psi) \wedge \neg\psi$, which is equivalent with $\mathbb{M}, t' \vDash T(\neg\psi)$, by ($T$). Thus, there is, and there is not, a t' such that tRt' and $\mathbb{M}, t' \vDash T(\neg\psi)$, which is impossible. ∎

$$\text{(d')} \quad \mathbb{M}, t \vDash T(\Diamond(\underline{\psi})) \rightarrow T(\neg\Box(\neg\psi)), \text{ for every } \mathbb{M}, t$$

Proof Consider an arbitrary time t in a model $\mathbb{M}$, and assume that $\mathbb{M}, t \vDash T(\Diamond(\underline{\psi}))$. We show that $\mathbb{M}, t \vDash T(\neg\Box(\neg\psi))$. As a first step, notice that since $\mathbb{M}, t \vDash T(\Diamond(\underline{\psi}))$, it is also the case that $\mathbb{M}, t' \vDash E\psi \rightarrow T\psi$, for some t' such that tRt', by (T), ($\wedge$), and ($\Diamond$). Since $\mathbb{M}, t' \vDash E\psi$ by (ERed), we obtain $\mathbb{M}, t' \vDash T\psi$ by *modus ponens*, which in turn implies $\mathbb{M}, t' \vDash \psi$, by ($T$) and ($\wedge$). Thus, $\mathbb{M}, t' \nvDash \neg\psi$, by ($\neg$), as well as $\mathbb{M}, t' \nvDash \neg T(\neg\psi)$, by ($T$). But $\mathbb{M}, t' \vDash E(\neg\psi)$, by (ERed), meaning $\mathbb{M}, t' \nvDash E(\neg\psi) \rightarrow T(\neg\psi)$, by ($\rightarrow$). By ($\Box$), it follows that $\mathbb{M}, t \vDash \neg\Box(\neg\psi)$. But $\mathbb{M}, t \vDash E(\neg\Box(\neg\psi))$, by (ERed), and thus also $\mathbb{M}, t \vDash E(\neg\Box(\neg\psi)) \wedge \neg\Box(\neg\psi)$, by ($\wedge$). This means that $\mathbb{M}, t \vDash T(\neg\Box(\neg\psi))$, by ($T$), which finishes the proof. ∎

$$\text{(d*')} \quad \mathbb{M}, t \vDash T(\neg\Box(\neg\psi)) \rightarrow T(\Diamond(\underline{\psi})), \text{ for every } \mathbb{M}, t$$

Proof Consider an arbitrary time t in a model $\mathbb{M}$, and assume, towards a contradiction, that $\mathbb{M}, t \vDash T(\neg\Box(\neg\psi))$ and yet $\mathbb{M}, t \vDash \neg T(\Diamond(\underline{\psi}))$. Notice, first, that since $\mathbb{M}, t \vDash T(\neg\Box(\neg\psi))$, it is also the case that $\mathbb{M}, t \vDash \neg\Box(\neg\psi)$, by ($T$) and ($\wedge$). Thus, by ($\Box$) and De Morgan, either (A) $\mathbb{M}, t \vDash \neg E(\neg\psi)$ or (B) there is a t' such that tRt' and $\mathbb{M}, t' \nvDash E(\neg\psi) \rightarrow T(\neg\psi)$. Since $\mathbb{M}, t \vDash E(\neg\psi)$ by (ERed), (A) is false, so (B) is true. By ($\rightarrow$) we obtain from (B) that there is a t' such that tRt' and $\mathbb{M}, t' \nvDash T(\neg\psi)$. By ($T$), ($\neg$) and De Morgan, this gives us $\mathbb{M}, t' \vDash \neg E(\neg\psi) \vee \neg\neg\psi$, for some t' such that tRt', which reduces to $\mathbb{M}, t' \vDash \neg\neg\psi$ since $\mathbb{M}, t' \vDash E(\neg\psi)$ by (ERed). But $\mathbb{M}, t' \vDash \neg\neg\psi$ just in case $\mathbb{M}, t' \vDash \psi$, by ($\neg$); and $\mathbb{M}, t' \vDash \psi$ can be extended into $\mathbb{M}, t' \vDash E\psi \wedge \psi$ by (ERed) and ($\wedge$). Thus, by (T), there is a t' such that tRt' and $\mathbb{M}, t' \vDash T\psi$. Notice, second, that by (T) and De Morgan $\mathbb{M}, t \vDash \neg T(\Diamond(\underline{\psi}))$ just in case $\mathbb{M}, t \vDash \neg E(\Diamond(\underline{\psi})) \vee \neg\Diamond(\underline{\psi})$, which reduces to $\mathbb{M}, t \vDash \neg\Diamond(\underline{\psi})$ since $\mathbb{M}, t \vDash E(\Diamond(\underline{\psi}))$ by (ERed). But by ($\Diamond$) and De Morgan this implies that either (C) $\mathbb{M}, t \vDash \neg E\psi$ or (D) there is no t' such that tRt' and $\mathbb{M}, t' \vDash E\psi \rightarrow T\psi$. Since $\mathbb{M}, t \vDash E\psi$ by (ERed), (C) is false, so (D) is true. But (D) is true just in case for all t' such that tRt', $\mathbb{M}, t' \nvDash E\psi \rightarrow T\psi$, which by ($\rightarrow$) implies that $\mathbb{M}, t' \nvDash T\psi$ for all t' such that tRt', or, equivalently, that there is no t' such that tRt' and $\mathbb{M}, t' \vDash T\psi$. Thus, there is, and there is not, a t' such that tRt' and $\mathbb{M}, t' \vDash T\psi$, which is impossible. ∎

(e') $\mathbb{M}, t \vDash T(\Box(\underline{\psi_1 \to \psi_2})) \to (T(\Box(\underline{\psi_1})) \to T(\Box(\underline{\psi_2})))$, for every $\mathbb{M}, t$

Proof Consider an arbitrary time t in a model $\mathbb{M}$, and assume that both $\mathbb{M}, t \vDash T(\Box(\underline{\psi_1 \to \psi_2}))$ and $\mathbb{M}, t \vDash T(\Box(\underline{\psi_1}))$. We show that $\mathbb{M}$ $t \vDash T(\Box(\underline{\psi_2}))$. Notice, first, that since $\mathbb{M}, t \vDash T(\Box(\underline{\psi_1 \to \psi_2}))$, it is also the case that $\mathbb{M}, t' \vDash E(\psi_1 \to \psi_2) \to T(\psi_1 \to \psi_2))$ for all t' such that tRt', by (T), ($\wedge$), and ($\Box$). Notice, second, that since $\mathbb{M}, t \vDash T(\Box(\underline{\psi_1}))$, it is also the case that $\mathbb{M}, t' \vDash E\psi_1 \to T\psi_1$ for all t' such that tRt', by parity of reasoning. Thus, both $\mathbb{M}, t' \vDash E(\psi_1 \to \psi_2) \to T(\psi_1 \to \psi_2))$ and $\mathbb{M}, t' \vDash E\psi_1 \to T\psi_1$, for all t' such that tRt' – call this (A). Now consider the following lemmas.

(L_1) $\mathbb{M}, t \vDash T(\varphi \to \psi) \to (T\varphi \to T\psi)$, for every $\mathbb{M}, t$.

Proof of (L_1) Consider an arbitrary time t in a model $\mathbb{M}$, and assume that both $\mathbb{M}, t \vDash T(\varphi \to \psi)$ and $\mathbb{M}, t \vDash T\varphi$. We show that $\mathbb{M}, t \vDash T\psi$. First, since $\mathbb{M}, t \vDash T(\varphi \to \psi)$, it is also the case that $\mathbb{M}, t \vDash \varphi \to \psi$, by ($T$) and ($\wedge$). By parity of reasoning, we obtain $\mathbb{M}, t \vDash \varphi$ from $\mathbb{M}, t \vDash T\varphi$. But $\mathbb{M}, t \vDash \varphi \to \psi$ and $\mathbb{M}, t \vDash \varphi$ jointly establish $\mathbb{M}, t \vDash \psi$, by *modus ponens*. Since $\mathbb{M}, t \vDash E\psi$, by (ERed), we obtain that $\mathbb{M}, t \vDash E\psi \wedge \psi$, by ($\wedge$). By ($T$), this is equivalent with $\mathbb{M}, t \vDash T\psi$, which finishes the proof of (L_1). ∎

(L_2) For every $\mathbb{M}, t$, if both $\mathbb{M}, t \vDash E(\varphi \to \psi) \to T(\varphi \to \psi)$ and $\mathbb{M}, t \vDash E\varphi \to T\varphi$, then also $\mathbb{M}, t \vDash E\psi \to T\psi$.

Proof of (L_2) Consider an arbitrary time t in a model $\mathbb{M}$, and assume that both $\mathbb{M}, t \vDash E(\varphi \to \psi) \to T(\varphi \to \psi)$ and $\mathbb{M}, t \vDash E\varphi \to T\varphi$. We show that $\mathbb{M}, t \vDash E\psi \to T\psi$. By (ERed), we have that $\mathbb{M}, t \vDash E\varphi$ and $\mathbb{M}, t \vDash E(\varphi \to \psi)$. Moreover, from $\mathbb{M}, t \vDash E(\varphi \to \psi)$ and $\mathbb{M}, t \vDash E(\varphi \to \psi) \to T(\varphi \to \psi)$ we obtain that $\mathbb{M}, t \vDash T(\varphi \to \psi)$, by *modus ponens*. But $\mathbb{M}, t \vDash E\varphi \to T\varphi$ and $\mathbb{M}, t \vDash E\varphi$ jointly imply $\mathbb{M}, t \vDash T\varphi$, again by *modus ponens*; and $\mathbb{M}, t \vDash T\varphi$ and $\mathbb{M}, t \vDash T(\varphi \to \psi)$ jointly imply $\mathbb{M}, t \vDash T\psi$, by ($L_1$). Thus, by ($\to$), it follows that $\mathbb{M}, t \vDash E\psi \to T\psi$, which finishes the proof of (L_2). ∎

From (A) and (L_2) it follows that $\mathbb{M}, t' \vDash E\psi_2 \to T\psi_2$, for all t' such that tRt' – call this (B). Also, we have that $\mathbb{M}, t \vDash E\psi_2$, by (ERed), which along with (B) implies $\mathbb{M}, t \vDash \Box(\underline{\psi_2})$, by ($\Box$). Since $\mathbb{M}, t \vDash E(\Box(\underline{\psi_2}))$ by (ERed), we obtain that $\mathbb{M}, t \vDash E(\Box(\underline{\psi_2})) \wedge \Box(\underline{\psi_2})$, by ($\wedge$). By ($T$), this is equivalent with $\mathbb{M}, t \vDash T(\Box(\underline{\psi_2}))$, which finishes the proof of (e'). ∎

Index

C. Geudens, L. Demey, *The Modal Logic of John Fabri of Valenciennes (c. 1500)*, SpringerBriefs in Philosophy, https://doi.org/10.1007/978-3-030-98802-9

The manufacturer's authorised representative in the EU is Springer Nature Customer Service Centre GmbH, Europaplatz 3, 69115 Heidelberg, Germany. If you have any concerns regarding our products, please contact ProductSafety@springernature.com

Printed and bound by CPI Group (UK) Ltd, Croydon, CR0 4YY
30/09/2026
02225468-0013